Medicine Demystified

(Volume 3)

From Taboo to Wellness: The Facts behind Menopause

Authored By

Professor Peter Hollands

Freelance Consultant Clinical Scientist
Huntingdon, Cambs PE261LB
UK

Medicine Demystified

(Volume 3)

From Taboo to Wellness: The Facts behind Menopause

Author: Peter Hollands

ISSN (Online): 2737-5064

ISSN (Print): 2737-5056

ISBN (Online): 978-981-5124-21-7

ISBN (Print): 978-981-5124-22-4

ISBN (Paperback): 978-981-5124-23-1

Published by Bentham Science Publishers Pte. Ltd. Singapore. All Rights Reserved.

First published in 2023.

need for a court order if at any point you breach any terms of this License Agreement. In no event will any delay or failure by Bentham Science Publishers in enforcing your compliance with this License Agreement constitute a waiver of any of its rights.

3. You acknowledge that you have read this License Agreement, and agree to be bound by its terms and conditions. To the extent that any other terms and conditions presented on any website of Bentham Science Publishers conflict with, or are inconsistent with, the terms and conditions set out in this License Agreement, you acknowledge that the terms and conditions set out in this License Agreement shall prevail.

Bentham Science Publishers Pte. Ltd.
80 Robinson Road #02-00
Singapore 068898
Singapore
Email: subscriptions@benthamscience.net

**BENTHAM
SCIENCE**

CONTENTS

PREFACE i
    CONSENT FOR PUBLICATION i
    CONFLICT OF INTEREST ii
    ACKNOWLEDGEMENT ii

DEDICATION iii

CHAPTER 1 A BIT OF HISTORY 1
    'THE CHANGE' 1
    INTRODUCTION 2
        The Four Stages of Menopause 3
            1. Premature Menopause or Premature Ovarian Insufficiency (POI) 3
            2. Perimenopause 3
            3. Menopause 7
            4. Postmenopause 7
    KEYPOINTS OF CHAPTER 1 7

CHAPTER 2 THE MENOPAUSAL WOMEN 8
    NEXT STEPS 9
    FEAR 9
    WHAT IS CAUSING FEAR? 9
    MYTHS 10
        Femininity and the Menopause 12
    WISHES OF MENOPAUSAL WOMEN 12
    PRAYERS OF THE MENOPAUSAL WOMAN 13
    KEY POINTS OF CHAPTER 2 13

CHAPTER 3 THE DIAGNOSIS AND STANDARD TREATMENT OF MENOPAUSE 15
    INTRODUCTION 15
    DIAGNOSIS OF MENOPAUSE 15
        'Do It Yourself' Menopause Diagnostic Kits 16
        Hormone Replacement Therapy (HRT) 17
        Methods of Taking HRT 18
        The Treatment Routines for HRT 20
        Cyclical HRT 20
        Continuous Combined HRT 21
        Side Effects of HRT 21
        Side Effects of Oestrogen in HRT 21
        Side Effects of Progestogen 22
        Weight Gain 22
        Potential Serious Side Effects of HRT 23
        Sleep and HRT 23
    CONCLUSION 24
    KEY POINTS OF CHAPTER 3 24

CHAPTER 4 THE ALTERNATIVE TREATMENTS OF MENOPAUSE 25
    INTRODUCTION 25
    LIFESTYLE 25
    EXERCISE 25
    DIET 26
    BE COOL 26
    FOOD 26

STRESS 27
SMOKING AND VAPING 27
VAGINAL DRYNESS 27
PHARMACEUTICALS 27
TIBOLONE 28
ANTIDEPRESSANTS 28
CLONIDINE 28
BIOIDENTICAL OR 'NATURAL' HORMONES 29
COMPLEMENTARY THERAPIES 29
HERBAL REMEDIES 30
ACUPUNCTURE IN THE TREATMENT OF MENOPAUSE 30
    Hot Flush and Night Sweat Treatment using Acupuncture 31
    Pain Treatment Using Acupuncture 31
    Mood Swings and Anxiety Treatment Using Acupuncture 32
    Insomnia Treatment Using Acupuncture 32
    Vaginal Dryness Treatment Using Acupuncture 32
PLANT BASED EXOSOMES 32
CONCLUSION 33
KEY POINTS OF CHAPTER 4 33

CHAPTER 5 THE MALE MENOPAUSE 34
INTRODUCTION 34
MALE AND FEMALE BIOLOGY 34
THE MALE ANDROPAUSE 35
MECHANISM OF THE MALE ANDROPAUSE 35
THE 'MID-LIFE CRISIS' 35
SYMPTOMS AND COMPLICATIONS OF THE ANDROPAUSE 36
DIAGNOSIS OF THE ANDROPAUSE 36
TREATMENT OF THE ANDROPAUSE 37
KEY POINTS OF CHAPTER 5 38

CHAPTER 6 REGENERATIVE MEDICINE AND THE MENOPAUSE 39
INTRODUCTION 39
STEM CELLS AND TREATMENT OF THE MENOPAUSE 40
    Very Small Embryonic-like (VSEL) Stem Cells In The Treatment Of The Menopause 41
    VSEL Stem Cells in the Human Body 42
    Pluripotent VSEL Stem Cells 42
    VSEL Stem Cells in the Blood of Everyone 43
    Activation of VSEL Stem Cells Derived from Circulating Blood 43
    How is all of this Relevant to The Menopause? 43
    How May the QiGen Protocol help in Menopause? 45
VSEL STEM CELL AGE 47
'YOUNG' DONOR VSEL STEM CELLS 47
HURDLES STILL TO CLEAR 48
CONCLUSION 49
KEY POINTS OF CHAPTER 6 49

CHAPTER 7 THE PSYCHOLOGY OF PREMATURE MENOPAUSE, PERIMENOPAUSE
AND MENOPAUSE 50
INTRODUCTION 50
'THE LUCKY ONES' 50
    The Psychology of Premature Menopause or Premature Ovarian Insufficiency (POI) 51

Mental Health 52
The Psychology of the Perimenopause and Menopause 52
Social Factors in Perimenopausal and Menopausal Depression 53
Psychological Traits and Perimenopausal and Menopausal Depression 54
Psychological Adversity as a Child 55
'Brain Fog' in the Perimenopause and Menopause 55
The Cause of Brain Fog 55
CONCLUSION 57
KEY POINTS OF CHAPTER 7 57

CHAPTER 8 THE MEDIA, CELEBRITIES AND MENOPAUSE 59
INTRODUCTION 59
THE MEDIA AND THE MENOPAUSE 60
Celebrities and The Menopause 61
KEY POINTS OF CHAPTER 8 63

CHAPTER 9 FAMILY, FRIENDS AND WORK 64
INTRODUCTION 64
FAMILY AND FRIENDS 64
The Role of the Husband or Male Partner of the Menopausal Woman 68
The Single Woman and The Menopause 71
NEW ACTIVITIES 71
WORK AND THE MENOPAUSE 72
KEY POINTS OF CHAPTER 9 73

CHAPTER 10 LESBIANS 74
INTRODUCTION 74
THE MENOPAUSE IN LESBIAN WOMEN 74
Associated Risks of the Menopause in Lesbian Women 75
Common Misunderstandings in the Sexual Health of Lesbians 76
Domestic Violence and Anxiety 76
Action Needed to Improve the Care of Lesbian Menopausal Women 77
CONCLUSION 77
PSYCHOLOGICAL STRENGTH 78
KEY POINTS OF CHAPTER 10 78

CHAPTER 11 ADVICE TO MENOPAUSAL PATIENTS 79
INTRODUCTION 79
THE FEMALE EXPERIENCE OF THE MENOPAUSE 80
Female and Male Education 81
Primary Care Health Providers 82
Hormone Replacement Therapy (HRT) 83
Anti-Depressants 84
Healthcare Professional Training 84
KEY POINTS OF CHAPTER 11 85

CHAPTER 12 A FINAL THOUGHT 86
INTRODUCTION 86
MENOPAUSE AND THE ANIMAL WORLD 86
SLIGHTLY MORE CONVINCING EXAMPLES OF FEMALE ANIMAL MENOPAUSE 87
THE GRANDMOTHER HYPOTHESIS 88
ELEPHANTS 88
COMPETITION FOR FOOD AND RESOURCES 89

    CONCLUSION 90

    KEY POINTS OF CHAPTER 12 90

USEFUL LINKS 91

    SUGGESTED FURTHER READING 92

GLOSSARY OF TERMS 93

SUBJECT INDEX 100

PREFACE

This book is inspired by my own experiences as a Clinical Embryologist helping patients through fertility treatments and as a Clinical Scientist helping patients through treatment using stem cell technology for various diseases. This book is not easy to read in many places. I do not apologise for this because to understand and cope with anything properly, the hard facts must be known, understood and accepted. This book is neither complicated nor does it use terminology which is unknown to the reader (there is also a comprehensive glossary). The book is tough to read because you may see statements and ideas which really hit home, and we all feel this at some point in our lives. Try to stick with it, it might be tough and even upsetting, but afterwards, you will hopefully feel empowered to manage your menopause and feel better. The more you learn, the more you can be pro-active in the treatment and management of your menopause instead of remaining a 'victim' of it.

The purpose of the 'Medicine Demystified' series of books (a total of 10 books) is to take difficult or complex areas of medicine and to provide clear, unambiguous and factual information on these subjects. Most patients feel vulnerable and frightened at some point in their medical treatment (whatever that might be) and are often 'too scared' to ask vital questions. If patients have a better understanding and can ask the right questions to the right people at the right time, then a lot of anxiety, stress and worry will be taken away. This 'empowerment' of patients is extremely important to me, and I have worked to achieve it throughout my whole career. A domineering 'don't ask me' healthcare professional is neither useful nor effective. We all know that many healthcare professionals operate under enormous stress. Despite this, a few moments taken to actually listen to the patient can be priceless to the eventual outcome for that patient.

My own experience in menopause comes not only from family members (I am sure that everyone knows someone who has been through menopause) but also from the thousands of fertility patients I have seen over the years. Fertility patients often first appear for treatment at fertility clinics when they are, in fact, starting, or even well into, the perimenopause. This is a 'double stress' situation. Not only are the patients dealing with perimenopause, but they are also dealing with the fact that their fertility is rapidly declining, and their chances of success with fertility treatment are declining by the day. There are two problems that result in this common situation:

1. Poor education about the perimenopause/menopause and its' implications

2. Poor understanding about the ongoing decline in female fertility with age and to the limits of current fertility treatments.

This book deals with problem number 1, and my second book in this series, 'The Fertility Promise, ' deals with every aspect of infertility and fertility treatment.

I have tried to be as factual, understanding, supportive and compassionate as possible throughout this book. I have talked about 'routine' matters relating to menopause and exciting, cutting-edge ideas which could one day even delay or 'reverse' menopause. I hope that you find the book useful, empowering, and most of all, enjoyable.

CONSENT FOR PUBLICATION

Not applicable.

CONFLICT OF INTEREST

The author declares no conflict of interest, financial or otherwise.

ACKNOWLEDGEMENT

Declared none.

Peter Hollands
Freelance Consultant Clinical Scientist
Huntingdon, Cambs PE261LB
UK

DEDICATION

This book is dedicated to my wife, Louise Barrett, for her love, dedication and support. I must also thank my cardiac surgeon Mr. Ian Wilson and everyone at Liverpool Heart and Chest Hospital, without whom none of this would be possible!

CHAPTER 1

A Bit of History

(An Overview of the Historical Development of the Recognition and Understanding of Menopause.)

"It is fairly brutal, and you go through all the accompanying side effects: hot flushes, weight gain, a sense of mourning for lost youth, sexiness, and somehow the point in anything. I became depressed, which I ended up getting help with."

Jennifer Saunders

'THE CHANGE'

Before we make a start on this book, I want to make it clear that there is no such thing as 'the change'. This is an expression often wrongly used instead of perimenopause/menopause, which probably dates back to the 1950's or even earlier when discussions of such medical matters were 'taboo' and often whispered about in quiet groups. There are, in fact, millions of 'changes' in the human body every few seconds. The brain processes and reacts to millions of electrical inputs, and the other organs in the body carry out millions of chemical reactions. Air moves in and out of the lungs to take in oxygen and remove carbon dioxide. The kidneys remove waste products and toxins. The liver carries out some of the most amazing biochemistry on Earth every second. The heart pumps blood through the body at three feet per second. The red cells in our blood fly through our veins, carrying oxygen to, and carbon dioxide away, from our organs and tissues. White cells in our blood fight infection and manufacture antibodies, *e.g.*, antibodies in response to the COVID-19 vaccine or any other type of vaccine. Platelets in our blood stop us from bleeding to death. Reproductive organs are constantly (up to the perimenopause) producing gametes which, when brought together, can form a new human! Stem cells replace and renew many tissue types in the body daily. The body we had a minute ago is different from the body that we have a minute later. Change is constant and natural; without it, we would all die in a few minutes. The living human body is not a static piece of tissue, it may seem so, but it most definitely is not. If you read around about perimenopause/menopause, you will no doubt find reference to 'the change' in some older references, and if you do, then please mentally delete it and replace it with perimenopause/menopause. You may also find reference to something called the 'climacteric'. This is the medical term for menopause and will not be used in this book because it will just lead to further confusion. Let's keep it simple, perimenopause and menopause are perimenopause and menopause. No more, no less.

INTRODUCTION

There are approximately 3.7 billion women on Earth, and around 1.9 billion of these women are of reproductive age (15-49 years old). This means that there are approximately 1.9 billion women who, in the next 34 years, will go into menopause. As the global population increases, so will the incidence and suffering related to menopause. Those women in developing countries with poor or even non-existent healthcare will suffer the most. Nevertheless, those women in highly developed countries are far from immune to the symptoms and suffering brought by the menopause. The term menopause was first introduced in 1821 by Dr. Charles Negrier, and he characterised it as 'depression, hot flushes and irregular periods to problems of the uterus.' Over two hundred years later, the problems associated with menopause still exist, and the interest to help women suffering from the menopause has, until very recent times, been of little interest to the medical profession. Other notable events in 1821 include the death of Emperor Napoleon I in exile on the island of St. Helena, the Coronation of George IV, and the United States taking possession of Florida from Spain.

It took another hundred years before scientists started to link the reduction in female hormone production to the symptoms of menopause. Following this, in 1942, the first commercial hormone treatment (made from pregnant mare urine) became available. This early hormonal therapy was unreliable and often prescribed only for a short time which meant that any benefit women might have enjoyed quickly disappeared. As a result, many women resorted to alcohol or even drugs, such as Laudanum, to try to block out the symptoms of desperate women.

It then took until 1963 (the year of the Great Train Robbery and the assassination of John F. Kennedy, amongst other things) for a serious medical publication to be written by Robert Wilson and his wife, which highlighted the plight of women suffering the symptoms of menopause. They demanded the development of better, safe, and more effective treatments for menopause. This resulted in the development of Hormone Replacement Therapy (HRT) which remains the mainstay of menopause treatment today. Nevertheless, HRT does not 'suit' all women, and other therapeutic approaches are constantly being assessed.

The Four Stages of Menopause

The following description of the different phases and symptoms of menopause may be very difficult reading for some people. If so, please accept my apologies at this stage, my aim is to be clear and helpful and not to add to the already considerable stress and confusion of menopause.

It is extremely important to clearly understand the different stages of menopause as these can often be the most worrying to most women because they fear other diseases and, in fact, what is happening is the menopause. The four stages of menopause are:

1. Premature Menopause or Premature Ovarian Insufficiency (POI)

Some women will undergo menopause at a young age, and this is known as premature menopause or premature ovarian insufficiency (POI). There are several possible causes of premature menopause (POI), such as:

- Hysterectomy (surgical removal of the uterus) in a young woman. This may be for a variety of reasons, but one consequence is the possibility of premature menopause.
- Oophorectomy (surgical removal of both of the ovaries) in a young woman for various reasons. In this case, the symptoms of menopause will begin immediately.
- Premature Ovarian Failure (POF). This may happen for unexplained or unknown reasons (known medically as idiopathic) in an otherwise healthy woman. POF may have a genetic base; it may be related to other abdominal surgery and will result from cancer treatments, such as radiation or chemotherapy. There is some evidence to suggest that some women may produce insufficient or abnormal follicles in their ovaries, resulting in poor quality or immature eggs which are unable to be fertilised.

Whatever the reason, this is a devastating, life-changing diagnosis for a young woman, and anyone feeling that they might be undergoing premature menopause should seek medical advice immediately.

2. Perimenopause

This can be a period of 3 to 5 years (it may, of course, be shorter or longer for some women) where the production of the female hormone called oestrogen starts to fall. On average, perimenopause starts when the woman is in her late 40's and results in many worrying symptoms, such as:

i. Hot flushes. These are episodes where the woman may feel extremely hot even though the ambient temperature is cool. Hot flushes can occur at any time and often pass within 3-5 minutes. This is a frightening sensation for most women and often causes unnecessary anxiety about more serious underlying disease. The fact is that hot flushes are inconvenient, embarrassing and stressful, but they do not pose an immediate medical emergency. There is also the psychological aspect of hot flushes, where the woman may feel afraid and anxious that a hot flush will coincide with an important event or meeting. The actual cause of hot flushes is currently not well understood, but it is thought to be due to hormonal changes acting on a part of the brain called the hypothalamus, where body temperature is controlled. Hot flushes may persist for 10 or more years which is a daunting thought!

ii. Insomnia. Many women will suffer some level of insomnia during perimenopause. These may be exacerbated by hot flushes occurring during sleep. This level of insomnia may have a major effect on the cognitive abilities of a menopausal woman and also induce the 'brain fog', which will be discussed later.

iii. Night Sweats. All of the symptoms of menopause are highly disturbing and worrying to woman, but night sweats are arguably the most disturbing. A night sweat is when a woman awakes in the night (sometimes every night) covered in sweat from head to toe which is often so severe that the bed is also wet from the sweat being produced. These night sweats are related to the hot flushes referred to above and result in a further disturbance of sleep which only exacerbates the overall symptoms of menopause.

iv. Elevated heart rate. Some menopausal women may experience an elevated heart rate. This can be a frightening experience because the woman fears that she may be having a heart attack. This results in more panic and anxiety, which develops into a vicious circle of symptoms and panic. The elevated heart rate is related to the hormonal changes happening during perimenopause and rarely, if at all, is an indication of heart disease. This does not mean that such a symptom should be ignored, but re-assurance should be sought from a healthcare professional.

v. Irritability, depression and anxiety. These are possibly the worst and potentially most damaging symptoms of perimenopause. Irritability will affect every aspect of the perimenopausal woman's daily life in terms of interactions with friends, family, work colleagues, and any other interactions which might normally happen during any day. This may result in the perimenopausal women becoming increasingly isolated with an increased feeling of loneliness. This can further develop into clinical depression, and if the perimenopausal woman does not seek help, then the results can be catastrophic and even fatal.

Any woman, perimenopausal or not, who feels that they are entering depression must seek help immediately. The common signs of depression are:

○ Feeling sad or 'empty'. These feelings often materialise and persist for many weeks, months or even years. You may experience a feeling of despair or total loneliness, which does not go away.

○ Hopeless and Helpless. This is the feeling that nothing is going to change or improve and that there is no one available who can help.

○ Worthless. A feeling of a total lack of self-worth and meaning in existence. This may be coupled with a feeling of being a 'burden' to other people, including even family members. Such people may also harbour suicidal thoughts.

○ Guilt. These feelings in a depressed person are inappropriate and disproportionate to the person but still feel very real to the depressed person. The guilt may be based on past or present events and can sap you of energy.

○ Loss of enjoyment. Everyone has something which they really enjoy, I like writing and using my knowledge to help people. Others may enjoy sports, social interactions, music or even sex! A person suffering from depression may lose all interest in these activities and will often strongly deny this lack of interest when challenged.

○ Anger and Irritability. These are common emotions in everyday life, but in depression, they may be amplified and controlling. They may also be linked to other symptoms, such as loss of sleep.

○ Tired. This may, of course, apply to anyone. We all feel tired at some point in our lives, and this may not be a sign of depression; we are simply tired! In terms of depression, tiredness may manifest as an inability to wake-up in the morning, general everyday fatigue and even a constant feeling of tiredness. This may result in difficulty with going to work, cooking or any 'routine' activity. This tiredness may, of course, be directly related to poor sleep or insomnia.

○ Insomnia. Some people going through the peri-menopause/ menopause may suffer from depression, and some may experience insomnia. This may manifest indifficulty getting to sleep or staying asleep. This may result in behaviour such as staying up late at night, waking up very early, or erratic and broken sleep, which will only tend to worsen insomnia. Insomnia can be extremely damaging to general physical and mental health. If you are experiencing any level of insomnia, then please seek medical advice.

○ Concentration, memory and decision-making. These are the so-called 'cognitive' abilities that happen almost automatically in everyday life. In a depressed person, decision-making and everyday choices may become difficult and even stressful. Memory may also decline (both past and recent), resulting in missed appointments and an inability to recall recent events or discussions. This can be extremely distressing and, in menopause, is often referred to as 'brain fog'.

- Appetite. A depressed person may experience a loss of appetite and the related weight loss which results. Interest in food may be lost, and fasting for long periods may be common.
- Over-eating and obesity. Depression may result in over-eating in some people who can easily lead to obesity. This may be further complicated by a lack of motivation to exercise, which then forms a vicious circle.
- General aches, pains and symptoms. We all suffer from these at some time, but in depression, they may be more dominant and include digestive disorders, headaches and other unexplained aches and pains.
- Suicidal thoughts. The thought of suicide (known medically as suicidal ideation) may result from suffering from depression. Self-harming may become common, as may actual suicide attempts and even death. Anyone with suicidal thoughts must get help immediately by seeing their General Practitioner (Family Doctor) or going directly to a hospital emergency department. **Do not delay**; your life may be at risk.

vi. Increased facial hair. Perimenopausal women very often develop increased and clearly visible facial hair (medical term hirsutism) in perimenopause. This can be extremely stressful to most women, and advice should be sought from a physician if the presence of facial hair becomes stressful or causes undue embarrassment.

My purpose in describing the signs and symptoms of depression in such graphic detail is that they often coincide with the signs and symptoms of perimenopause. The result of this is that some women may be suffering from perimenopause and receive treatment for depression. The treatment for depression may alleviate some of the symptoms, but it will not change the underlying cause, which is perimenopause.

i. Vaginal dryness. This can be caused by a variety of reasons, including peri-menopause. It is uncomfortable and often painful and, of course, makes sexual intercourse a very unattractive prospect. This may, in turn, impact long long-standing relationships, cause further anxiety and worsen the whole situation. There are hormonal treatments, or even simple lubricants, which can resolve the problem and any woman with this symptom should see a physician as soon as possible.

ii. Urinary problems. There may be an increase in urinary infections, difficulty passing urine, cystitis and incontinence associated with perimenopause. There are also many other possible causes of these symptoms. If you suffer from any form of urinary problem, please see a physician as soon as possible. Please do not suffer in silence.

3. Menopause

The average age for a woman to enter menopause (assuming that all other factors are normal) is 51-52. The technical definition of menopause refers to 12 consecutive missed periods without underlying causes such as pregnancy and breastfeeding, illness and medication. It may take 1-3 years to go from perimenopause to menopause and postmenopause, but everyone is different, and symptoms will vary widely. Generalisations are best avoided when talking about menopause because this only adds to unnecessary expectations and worries.

4. Postmenopause

This is defined as one year after the last menstrual cycle. The symptoms suffered in perimenopause may continue in postmenopause, but it is inadvisable to make generalisations. Perhaps the most important physiological change during postmenopause is the decrease in the hormone oestrogen. This can increase the risk of heart disease and osteoporosis (bone density loss), which can have serious consequences later in life.

The purpose of this detailed introduction to the subject of menopause is not to frighten anyone but to explain clearly what may or may not happen during menopause. Each woman is different, and the experience of menopause by each woman is different. Some women 'sail through' the menopause with little or no discomfort or concerns, others suffer minor symptoms which are easily managed, but the unlucky ones will find themselves in the centre of a life-changing process over which they seem to have no control.

KEYPOINTS OF CHAPTER 1

- The human body is in constant change, literally every second. The term 'the change' for the menopause is incorrect and irrelevant.
- There are four stages to the menopause, each of which must be understood to reduce stress and anxiety.
- Be prepared to understand the symptoms of the perimenopause/menopause. This will reduce unnecessary anxiety and stress.
- There are many symptoms associated with the menopause, and the experience of each woman will be *different*. Understanding these symptoms and their variability is an important starting point for the overall understanding of the menopause.

CHAPTER 2

The Menopausal Women

(The Fears, Hopes, Wishes and Prayers of Menopausal Women)

The freedom of patient speech is necessary if the doctor is to get clues about the medical enigma before him. If the patient is inhibited, cut off prematurely, or constrained into one path of discussion, then the doctor may not be told something vital. Observers have noticed that, on average, physicians interrupt patients within eighteen seconds of when they begin telling their stories.

Jerome Groopman

It is perhaps a silly understatement to say that menopausal women worry. The reason for this is not that menopausal women are in some way different, but they are showing a normal response common to all humans. If someone gets blood in the urine, cannot shift a persistent cough, or has a headache that will just not go away, then everyone will worry. Any unusual symptoms suffered by a male or a female will almost certainly result in worry. This worry will often result in sleepless nights, and as the dreaded 'consultation' with a physician looms up, the worry increases even more. The patient worries that they will be told that they have a few weeks to live, or that major surgery is needed, or that a long, painful treatment regime will have to start, which may fail. They may, of course, be told that taking a course of antibiotics or undergoing some physiotherapy will cure the problem. Whatever the outcome, worry always precedes the diagnosis and treatment; it is human nature.

In the case of the perimenopausal woman, it is true to say that this worry may be even more severe. Is it cancer? Will I see my grandchildren grow up? Is it Alzheimer's disease? The symptoms are so wide-ranging and can be so intense that any eventuality could be imagined.

The first port of call for a perimenopausal woman is most often the General Practitioner (GP) or Family Physician (FP). This is a great first step, but of course these physicians do not specialise in menopause and may even just pick up on the anxiety and depression side of menopause and prescribe anti-depressants. Such an approach may help with the depression and anxiety being experienced but will have no great impact on the menopause itself.

NEXT STEPS

The next step may be a referral to an endocrinologist who specialises in diseases of the glands in the body and the hormones they produce, *e.g.*, the pancreas and diabetes. Such a physician will have excellent knowledge within the field of endocrinology. Menopause is caused by the declining function of the ovaries, which are certainly endocrine glands producing hormones. The endocrinologist will therefore pinpoint the diagnosis of perimenopause or menopause very easily, and this may be of some comfort to the patient who now has a clear diagnosis and perhaps can see a way forward. The treatments which may follow for menopause are described in subsequent chapters, but the basic premise of all current treatments is to simply 'replace' the missing hormones and therefore remove the debilitating symptoms of menopause. This works for some patients, but it may not be so successful for others.

FEAR

Some women going through perimenopause and menopause can develop levels of fear which can be life-changing in some cases. These fears can be many and varied, and once again, no generalisations can be made. Despite this, there have been reports of women who fear socialising, leaving the house, or even just meeting new people. There may be work pressures, such as deadlines or targets, which for the menopausal woman, can transform into all-encompassing fear. A very common fear is a fear of driving. Very competent and experienced women drivers may develop this fear and it might be related to the fear of losing concentration (perhaps due to 'brain fog') and potentially crashing or hurting someone else. In the extreme, the fears may become deep-seated and irrational, resulting in a block between the fear and reality. This can be extremely frightening for anyone but especially an already vulnerable woman who has previously been confident and outgoing. The fears may extend to the family members, whereas the menopausal woman may honestly fear that something bad is going to happen to a family member. This fear can become disruptive and all-consuming for some menopausal women but may not even be noticed by other women.

WHAT IS CAUSING FEAR?

The key cause of fear in menopause is due to low oestrogen levels, which in turn cause significant changes in the brain and in how we think and rationalise on a daily basis. Prior to menopause, this is 'automatic,' and governed by oestrogen, but when oestrogen control is lost, then the worst aspects of fear can arise. The part of the brain which is regulated by oestrogen controls, such things as survival, food, life and reproduction. This means that when oestrogen levels fall, the menopausal woman may experience fears of imminent danger, a threat to life, a lack of food or an inability to obtain food, and the fear of being no longer able to reproduce. These are very basic fears which threaten life and survival and are deeply set within each persons' consciousness. When these fears are amplified by the lack of oestrogen in menopause, the results can be very frightening and seem very real.

MYTHS

There are many myths surrounding menopause which can, in some cases, increase the levels of anxiety. There are approximately 13 million in the UK alone suffering from menopause, so clear, unbiased and most importantly, true information will help literally millions of people in the UK alone; globally, this will be billions. The following are common myths and must be ignored by anyone approaching or going through menopause:

- Menopause will happen at 50. This is a myth because menopause is a natural ageing process that can take many years and does not happen when the candles are blown out on the 50th birthday cake! Indeed, the average age when a woman will reach the menopause in the UK is 51. Despite this, menopause is a complex and slow process that usually occurs across a wide age range, such as 45-55. It must also be remembered that there is also the occurrence (about 1 in 100 women) of premature menopause, usually defined as menopause occurring before the age of 40. The basic message here is that women should not fear that menopause will one-day 'switch on', it is a long process with many stages, as described in Chapter 1.
- Menopause is a single event. This is clearly a myth. Menopause does not just 'happen'. The stages of menopause have been carefully described in Chapter 1 and clearly show that menopause is a long process (in everyone) that may take place over many years. There may be certain 'events' within the menopause process that may seem significant, but menopause itself is not a single event; it is a complex combination of many physiological changes.
- The experience of menopause is the same for all women. This is partially true because women may experience similar symptoms when going through menopause, but in physiological terms, the reaction of the body to menopause will vary greatly from woman to woman. Oestrogen levels reduce in menopause, but many other hormones, such as progesterone and testosterone, also decline, resulting in very complex physiological changes which cannot be standardised. It might be surprising to see that testosterone (the classic male hormone) is important in female physiology. Testosterone, at levels much lower than those in the male, is, in fact, very important in female physiology in terms of the

promotion of bone growth and strength, promotion of cognitive health (*e.g.*, the prevention of neurodegenerative disease, such as Alzheimer's disease) and in the female sex drive. This nicely illustrates the complex nature of endocrinology and the interactions of hormones in both male and female bodies.

- Menopause will only last a few years. This is an over-generalisation that may confuse and upset many women. On average, menopause symptoms last for approximately 4 years but 1 in 10 women may suffer symptoms for up to 12 years. This is not meant to scare or worry menopausal women but just ask them to keep in mind that there is no 'standard' in terms of the duration of menopausal symptoms, everyone is different, and any 'drift' from what is suggested to be normal should be dismissed as a myth.

- Weight gain is unavoidable in menopause. This, again, is partially true, in that, menopausal women may gain 4-7lbs (on average) during menopause due to changing hormone levels, which directly impact metabolism (the chemical reactions constantly taking place in the body) and in the re-distribution of fat around the body. The solution to this problem is to ensure that good fitness levels should be maintained by regular exercise (this may just be a 30-minute walk per day, you don't have to run marathons!) and also a healthy diet. A healthy diet means avoiding fatty processed foods and eating plenty of protein, fruit, vegetables, whole grains, dairy, and some fat. They key to a healthy diet is moderation and portion control, some chocolates, or 1-2 glasses of wine per week will do no harm.

- No treatment is available for the symptoms of menopause. This is simply incorrect. Patients must first seek help; suffering in silence will not make things get any better. Most people understand that Hormone Replacement Therapy (HRT) is a possible treatment of the menopause, but there is much more than just HRT available as a supporting or alternative treatment, which will be described later in this book.

- HRT has health risks. This will be explored further later in this book, but from the point of view of being a myth, it is true that there are some HRT has health risks for some women, but these risks are, in general terms, very low. These health risks, compared to benefits, will be discussed in detail later in this book.

- Discussion of menopause is a bad thing. This is very untrue. Healthcare professionals, patients, scientists, politicians, and the media must become more menopause aware. This is because better awareness will ultimately result in better treatment for menopausal women, and this is what we all want.

- This discussion has been enhanced considerably recently by people, such as Davina McCall, Michelle Obama, Oprah Winfrey, Dame Judi Dench, and many others talking publicly about their experiences of the menopause. These people are, of course, all fantastic communicators and advocates, and their input to the overall discussion is critical. My criticism, at the time of writing, is that there are few males who join the menopause debate. It is men who need to understand the menopause as much as women so that they understand and properly support their wives or partners when they go through menopause. Most men will feel extremely uncomfortable about even talking about the menopause, and for the sake of the women in the world, this has to change.

Femininity and the Menopause

Some women will feel that the menopause makes them in some way 'less feminine'. This is patently wrong, but nevertheless, it is a feeling which is expressed by some menopausal women. It is true that a menopausal woman can no longer get pregnant and have a baby, this is a biological fact with which we cannot really argue. This may be at the heart of the feeling of a lack of feminity. Despite this, the menopausal woman is still a living, breathing human being with the same or even amplified emotions, which cannot be ignored. Menopause can affect the way that sex feels and is enjoyed, and also, the libido or 'sex drive' can be reduced. The idea that a menopausal woman cannot have an orgasm is totally untrue; nevertheless, she may have less desire to achieve an orgasm. There are also physical changes that may make orgasm difficult to achieve an orgasm, such as vaginal dryness, irritation or tightness, leakage of urine, less clitoral sensitivity and pain during sex.

The menopause impacts primarily the reproductive ability of a woman. This has to be accepted as part of life. The rest of the life of a menopausal woman need not change, including sexual activity.

WISHES OF MENOPAUSAL WOMEN

I suppose that the ultimate wish of any menopausal woman is that either menopause would never happen or that there was a simple and effective treatment for everyone. Sadly, as we stand, neither of these can be fully achieved. Menopause is a natural part of the ageing process, and we will never be able to eliminate it unless our understanding and technology develop beyond all recognition in the future. The concept of anti-ageing is being debated and researched intensely at present, and one of the benefits of such work may be to manage the menopause more effectively. This will take a lot of time and may even be undesirable to apply as the menopause is part of nature and arguably something with which we should not interfere. The second wish, regarding simple and effective treatment for all, is also either unlikely or even impossible. A different philosophy is needed. We cannot stop the menopause, but we can manage it. We can offer current treatment, and the future will hold new ideas and technology which may manage it even better. We can educate and inform so that employers, partners, husbands and society, in general, can freely talk about menopause and better support the menopausal woman.

PRAYERS OF THE MENOPAUSAL WOMAN

Prayer is an intensely personal matter. I know nothing about theology or religion and, in fact, am an atheist, but this does not mean that I do not understand or support those people who feel that prayer is helpful and important. Some people may not believe in prayer, and others may pray daily. My experience is that when a person or a loved one falls ill for any reason, then prayer may give solace and hope to what might seem an impossible situation. Whether one believes in God or not, in times of stress, illness and conflict, prayer becomes important, and sometimes essential, to bear the pain and suffering. What is clear is that menopause is not a terminal disease, it is a natural bodily process, and as such, it might not invoke the need for prayer as other more destructive life events. Nevertheless, menopause can be a very traumatic event for some women, and this is where prayer may be helpful. If I prayed, my prayer would be for all people suffering throughout the world, be it because of famine, war, oppression, religion or illness, and for medical science to find cures for all diseases, which can be made available to the whole world to reduce suffering and pain. Your prayer may be very different from mine (I actually do not pray or believe in a higher power), but perhaps the intention might be the same: peace and health to all humanity.

KEY POINTS OF CHAPTER 2

- If you suspect the perimenopause or menopause is happening to you, then do not just sit and worry. Be pro-active, take action and consult a healthcare professional.
- Menopause can be diagnosed by simply measuring the hormone levels in the blood, but the correct management of menopause can only begin after this diagnosis. The diagnosis is not the end of the road; it is the start of the road.
- Fear will no doubt be part of your experience but try to manage it appropriately. Menopause is not a death sentence; it is a natural biological process that if managed correctly, might even be an 'easy' process.
- There are many myths surrounding the menopause, and 99.99% of them are nonsense. Try to understand these myths for your own peace of mind but do not dwell on them.
- Your femininity does not change because you have undergone the menopause. You can no longer have a period (a great relief to many women), and you can no longer have a baby. Nevertheless, you are still the same person you were before the menopause. Enjoy this phase of your life, do not fear it.

- If you feel that prayer and spirituality are important in your life, then embrace them and use them to optimise this second phase of your life.

CHAPTER 3

The Diagnosis and Standard Treatment of Menopause

(Hormone Replacement Therapy, HRT)

The 'Science' behind hormone replacement therapy has put women on a medically engineered, press-fueled, big pharma-funded roller coaster.

Willow Bay

INTRODUCTION

In this chapter, I will explain firstly how the menopause is diagnosed (this is more complicated than you might think) and then the standard treatment of the menopause, which is Hormone Replacement Therapy (HRT). It is most important for any woman going through the menopause to understand both the diagnosis and standard treatment of the menopause. This will help to reduce the anxiety often felt during the diagnosis phase and the medication she may be asked to take.

DIAGNOSIS OF MENOPAUSE

It is important to remember that menopause can happen across a very wide age range, from aged 40 to 60 years. For this reason, it is essential that you consult a GP or Family Physician about your symptoms as soon as they arise. This will ensure that any other diseases are excluded in the diagnostic process to come to a clear diagnosis of the menopause. The healthcare professional will be able to assess you as a patient and prescribe Hormone Replacement Therapy (HRT) if it is safe and effective to do so. Please note that HRT is not appropriate for everyone. The diagnosis of the menopause (or perimenopause) is most commonly based on your age, your symptoms (*e.g.,* hot flushes, night sweats as described in Chapter 1), and how often you have periods. Most women are unlikely to need further tests. The diagnosis can be more difficult if you are already taking any hormonal medication, which is often used in the treatment of heavy periods.

A blood test to assess the levels of Follicle Stimulating Hormone (FSH), which will be requested by your physician, may be required if:

- You are aged between 40 and 45 and have menopausal symptoms along with changes in the regularity of your periods.
- You are aged under 40 and there may be a possibility of premature menopause or premature ovarian insufficiency (POI).

FSH in the blood is at higher levels when a woman is in the menopause. Nevertheless, caution, knowledge and experience must be used by the healthcare professional in interpreting the levels of FSH in the blood, especially if:

- You are about to ovulate.
- You are taking a contraceptive containing progestogen, oestrogen or high-dose progestogen.

The reasons for this are that FSH rises just before normal ovulation, and the contraceptives may cause changes in your natural FSH levels. Both of these may easily be misinterpreted as the menopause by unwary healthcare professionals.

'Do It Yourself' Menopause Diagnostic Kits

At the time of writing, an 'over-the-counter menopause diagnostic kit' was available in the UK and USA. These can be purchased and used without the supervision of a physician. These menopause 'diagnostic' test kits will only give a qualitative estimate of your current FSH level in your urine at the time of the test. In contrast, the FSH blood test, ordered by a physician, gives an accurate quantitative FSH result. The qualitative 'menopause diagnostic kit' will, therefore, only indicate if your FSH levels are raised. There are numerous causes of this, including natural ovulation! The risk, therefore, is that if these over-the-counter tests are carried out without the supervision of a physician, then the interpretation of the test result may be completely wrong.

The accuracy of the over-the-counter' menopause diagnostic kits is agreed to be around 9 out of 10 (90%). This sounds good, but it still means that 10% of the test results are meaningless. That could mean many people receiving incorrect results. The accuracy of the FSH blood test is 100%.

It must also be remembered that as you grow older, your FSH levels may fluctuate both up and down during your menstrual cycle. During this time, the ovaries will continue to release eggs, and pregnancy is still possible. The accuracy of 'over-the counter' menopause tests is also reduced (meaning that false negatives may be obtained) by the following factors:

- Whether or not the test was carried out on the first-morning urine.
- Whether or not you drank large amounts (2-3L) of water before carrying out the test. This dilutes the urine, which in turn dilutes the level of FSH in the urine.
- Whether or not you use, or have recently stopped using, oral or patch contraceptives, HRT, or oestrogen supplements.

If you carry out an 'over-the-counter' menopause test and get a 'positive' result, then this means that you may be in a stage of the menopause. The next step, especially if you have any menopausal symptoms, is to see your GP or Family Physician. It is absolutely essential that you do not stop taking any contraceptives based on the result of a 'positive over-the-counter' menopause test. Doing so may risk an unwanted pregnancy.

If you carry out an 'over-the-counter' menopause test and get a 'negative' result, but you still have menopausal symptoms, then you still may well be in the perimenopause or menopause. Please do not assume that you have not reached the menopause because there could be many other reasons for a negative result, as described above. The key message here is that if you choose to use an 'over-the counter' menopause testing kit, then do not assume that the result it gives confirms or refutes whether or not you have the menopause. Simply put, these 'over-the-counter' tests are effectively meaningless and a waste of money. If you have menopausal symptoms and are worried about consulting a physician, this is the only safe and effective way to proceed. One final note on the 'over-th--counter' menopause tests is that they must not be used to assess your fertility as, once again, an unwanted pregnancy may result.

Hormone Replacement Therapy (HRT)

The concept behind HRT is to replace the hormones which are no longer being produced during the menopause so that these hormones will reduce or remove the symptoms of menopause. On first sight, this makes sense, but it is not as simple as that.

There are 2 main hormones used in HRT, and these are:

1. Oestrogen, which may include oestradiol, oestrone and oestriol.

2. Progestogens are synthetic versions of progesterone (*e.g.,* medroxypro-gesterone, dydrogesterone, levonorgestrel norethisterone). There is also a version known as micronised progesterone (sometimes called body identical or natural) that is chemically identical to the human hormone. The choice of which hormones and which combinations are suitable for you, will be made by your healthcare professional.

One very common question asked by many menopausal women is that why does HRT contain progestogen when the natural hormone is progesterone? It is very true that progesterone is one of the natural female sex hormones along with oestrogen. The production of progesterone from the ovaries declines during the peri-menopause and stops completely during the menopause. The answer to the question posed is that progestogens are synthetic forms of the naturally occurring hormone progesterone, and are therefore easier to manufacture than trying to collect 'natural' progesterone.

In a pre-menopausal woman, progesterone helps to develop and maintain the lining of the womb (uterus) and helps with the general quality of sleep. The use of progestogens in HRT is preferable to pure progesterone because progestogens do not carry the same blood clotting risk as progesterone and also may minimise any increased risk of breast cancer. It is possible to administer some progestogens using an intra-uterine coil. This achieves only local distribution of the medication to the uterus, with very little, if any, of the medication finding its' way to the rest of the body. This may be a great benefit to some women. Every medication comes with its' associated risks, and in the case of progestogens, this depends on the dosage and the different modes of delivery to the patient. There are many common side effects associated with progestogen therapy, including acne, oily skin, mood changes, weight gain, irregular bleeding, breast tenderness and constipation. It is extremely important to speak to your healthcare professional if any such side effects become distressing, as a change in medication or delivery route might solve the problem.

Some women may develop what is known as 'progesterone intolerance', which produces symptoms similar to pre-menstrual syndrome. If this becomes a problem, it can sometimes be resolved by using natural progesterone instead of progestogen. In this case, it is important to use the minimum required dose, and patients need close monitoring of the uterine lining. Finally, and I am probably repeating myself, it is important to remember that every woman's experience of the menopause and the reactions to HRT medication are unique to each woman.

Most women will take both hormones for their HRT. Those women who have had their uterus (womb) removed but the ovaries are still present will probably be prescribed oestrogen only for their HRT. It must be remembered that HRT is not a 'one size fits all' treatment and that variations in the content and method of taking HRT will vary considerably from woman to woman in order to achieve the safest and most effective medication for each person.

Methods of Taking HRT

There are 6 main methods by which HRT may be administered, these are:

1. Tablets. This is perhaps the most common way of taking HRT, and the tablets need only be taken once a day, so it is relatively easy to introduce this into a daily routine and not forget to take the tablets. Oestrogen only or oestrogen and progestogen can be prescribed using tablets. There is a slight note of caution in that the medical literature reports a slightly higher risk of blood clots (usually in the lungs, brain or legs) in patients who take HRT in tablet form. This is arguably a minor risk when compared with the benefits of HRT.

2. Skin Patches. This is a widespread and popular way of delivering HRT to menopausal women. The adhesive patches are applied directly to the skin, and the medication they contain crosses the skin and into the blood of the patient. Such patches need to be renewed every few days. The application of HRT medication directly to the skin may reduce some of the side effects of HRT by reducing, such things as indigestion resulting from taking tablets. The use of skin patches also seems to exclude the increased risk of blood clots seen in those patients taking tablets. HRT patches are arguably the most convenient and the best-tolerated form of HRT.

3. Oestrogen Gel. This is a gel, which is rubbed into the skin once a day and is absorbed in a similar way as the medication in patches. Once again, the use of oestrogen gel does not seem to increase the risk of blood clots. In those patients who still have a uterus (womb), some form of progestogen will also be prescribed in order to reduce the risk of uterine cancer.

4. Implants. These are small 'pellet-like' preparations of HRT, which are placed under the skin (usually in the abdominal area) by a physician and with the use of a local anaesthetic to mask any pain or discomfort related to inserting the implant. The implant will slowly release oestrogen and will continue to do so for several months until a replacement is required. The method of delivery of HRT is popular with some women who do not want to worry about taking medication in tablet form, patches or gels. If the uterus is still present, then the patient will also need to take a progestogen. An alternative way to deliver the progestogen is an intruterine system (IUS). A IUS releases progestogen hormone directly into the uterus and can do so for up to 3-5 years. Implant use is not common, and implants may not be available in some countries, but they are a good form of delivering HRT for some women.

5. Vaginal oestrogen. This is in the form of a cream or pessary placed inside the vagina. This approach may be effective in reducing the occurrence of vaginal dryness, but it will not relieve the other symptoms of menopause because the hormone delivered in the way does not travel throughout the body; it tends to stay in the vagina. Vaginal oestrogen does not carry the increased risks described

above and does not increase the risk of breast cancer. Even if the patient still has her uterus, then vaginal oestrogen can be used without the need to also take a progestogen.

6. Testosterone gel. It may sound a little strange to include the male hormone testosterone as a form of female HRT, but as described earlier, it does play a critical role in the female, especially in the increase of libido. Testosterone would be prescribed alongside the HRT described above if the woman had a very low sex drive, and this was causing her stress and concern. The dose of testosterone must be carefully monitored to avoid unwanted side effects such as acne and unwanted hair growth. We do not want to create bearded ladies!

The Treatment Routines for HRT

In the ideal world, every menopausal patient would have a treatment routine specifically designed for that patient. At the time of writing, this level of 'personalised' medicine has yet to arrive, but in the future, it will almost certainly become a possibility and may make the overall treatment of the menopause more effective. Only time will tell. At present, there are two main categories of HRT treatment routines:

1. Cyclical HRT

2. Continuous Combined HRT

Cyclical HRT

Cyclical HRT (sometimes known as sequential HRT) is usually prescribed to those women who have menopausal symptoms, as described earlier, but are still having periods. There are two possible routines when delivering cyclical HRT which are:

1. Monthly HRT. Here the woman is prescribed oestrogen to be taken every day and progestogen alongside oestrogen for the last 14 days of her menstrual cycle. This approach mimics what these hormones do naturally from puberty to menopause. Monthly HRT is found to be most effective in those women who are still having regular periods.

2. 3-monthly HRT. Here the woman is prescribed oestrogen to be taken every day and progestogen to be taken alongside oestrogen for 14 days before the end of the 3 month cycle. The 3-monthly HRT routine is usually prescribed for women with irregular periods. Women who take HRT using the 3-monthly routine should have a period every 3 months.

Continuous Combined HRT

Continuous combined HRT is usually prescribed for women who are post-menopausal. The definition of post-menopausal in this context is any woman who has not had a period for 1 year and, of course, where there are no other reasons why the periods have stopped, *e.g.,* removal of the ovaries or some types of chemotherapy for cancer. In Continuous Combined HRT, the woman has been prescribed both oestrogen and progestogen to be taken together every day for the rest of her life.

Side Effects of HRT

Every prescribed medication has potential side effects; HRT is no exception. If side effects are experienced when taking HRT (and not everyone gets side effects or the same side effects), then there is the possibility that the side effects will resolve with time. This, however, is a matter of perseverance because it may take up to 3 months for side effects to stabilise and resolve. Any woman who suffers from side effects from HRT, which persist for more than 3 months, should seek further medical advice. If the side effects are extreme or unbearable, then medical advice must be sought immediately, and women should not suffer for 3 months before seeking help.

As described above, there are 2 main components of HRT, which are oestrogen and progestogen. It is necessary to consider the side effects of these two components separately.

Side Effects of Oestrogen in HRT

The side effects of taking oestrogen as part of HRT may include:

- Breast swelling or tenderness
- Swelling in other parts of the body, *e.g.,* ankles
- Bloating, indigestion and feeling sick (nausea)
- Cramp in the legs (especially the calf muscles at night)
- Persistent headaches
- Vaginal bleeding

These side effects may be mild, or they may be over-powering. If you suffer side effects that you cannot tolerate or manage, it is important to seek medical advice immediately. Some women find that taking oestrogen tablets with food may reduce the feeling of sickness or indigestion. A low-fat and high-carbohydrate diet may help to reduce any breast tenderness which may be experienced. Regular exercise and stretching exercises, which are good things to do anyway, may help with leg cramps. If the woman is overweight based on Body Mass Index (BMI), which should be between 18.5 and 24.9 to be considered in the 'normal range', then weight loss may also help. A BMI of 30 or above is considered to indicate obesity and should be a matter of serious concern for the patient. It should be noted that Body Mass Index has to be interpreted very carefully as there are many exceptions to the general range (including post-menopausal women!). Once again, if BMI is of concern to you, then discuss it with a physician. In terms of moderating or reducing side-effects, it may be possible to prescribe oestrogen as a patch rather than a tablet, change the formulation of the oestrogen being taken or lower the dose of oestrogen taken. Your physician will explore all of these options if needed.

Side Effects of Progestogen

The side effects of progestogen are similar to those of oestrogen and can include:

- Headaches and sometimes migraine
- Breast tenderness
- Body swelling as for oestrogen
- Mood swings and, in extreme cases, depression
- Abdominal pain
- Acne
- Back pain
- Vaginal bleeding

Once again, these side effects may be managed very well by most women but if the side effects are extreme or unbearable, please seek medical advice immediately. Do not suffer in silence. If a feeling of depression develops (loss of interest in activities, events or just feeling 'low'), then this needs urgent medical attention. The change of medication type and dose also applies to progestogen as a method of controlling or removing side-effects.

Weight Gain

Many women worry about the possibility of gaining weight as a result of taking HRT. There is, at present, no evidence to support these worries, and weight gain should not be considered an inevitable consequence of taking HRT. Despite this, there is some evidence about natural weight gain in the menopause, even in those women who do not take HRT. This, therefore, appears to be a natural process and not related specifically to HRT. In the event of any significant weight gain, then it is important to take some form of regular exercise and eat a healthy diet with as few processed foods as possible. If you still feel that your weight is increasing, then take medical advice, and once again, do not suffer in silence.

Potential Serious Side Effects of HRT

There are more serious but relatively rare side effects of HRT, which are an increased risk of certain types of breast cancer and blood clots. In terms of breast cancer, there is little, if any, increase in risk for those women who take oestrogen-only HRT. The risk of breast cancer increases slightly in those women taking combined HRT. The risk is related to the length of time over which HRT is taken, and the risk falls back to normal levels if HRT treatment is discontinued. Since there may be a slight increase in breast cancer in those women taking HRT, it is very important to attend regular breast cancer screening as early detection of cancer always results in a better treatment outcome. If you have a particular concern about HRT and breast cancer, then please discuss it with a physician.

In terms of an increased risk of blood clots (which usually arise in the legs and may travel to the lungs where more serious damage may occur), there is no increased risk in those women who take HRT in the form of gels or patches. Those women who take HRT in tablet form are at a slightly higher risk of suffering blood clots. It is, therefore, important to be aware of any possible signs of a blood clot, such as pain in the legs or shortness of breath (or both), and to seek medical advice immediately if these symptoms appear.

Sleep and HRT

Perhaps one of the most debilitating symptoms of the menopause is a lack of sleep due to hot flushes, night sweats and insomnia. A study in 2022 suggested that oestradiol HRT improved sleep quality overall. The same study suggested that another form of oestradiol, known as oestradiol valerate (present in many HRT medications), did not improve the quality of sleep. Of particular interest was the conclusion that patch-delivered HRT was more effective in the improvement of the quality of sleep than HRT taken orally in tablet form. There are two things to note about these observations:

1. Everyone is different, and these broad generalisations may not apply to everyone. They are a good indication of perhaps where to start reviewing HRT if sleep is a particular problem.

2. There is no 'one size fits all' in medication and especially in areas such as HRT medication. If you are taking HRT and you are not feeling the benefits that you were expecting, then please see your physician to review the HRT. A simple cha-

nge may make a big difference to the quality of your life, but if you suffer in silence, the no-one benefits.

CONCLUSION

It is re-assuring that HRT does not significantly increase the risk of heart disease or stroke; when HRT is started before age 60, it may even reduce the risk of heart disease or stroke. As with all medication, there has to be a balance between risk and benefit. In the case of HRT, it is generally agreed that the benefits considerably outweigh the risks, but at the same time, it is important to be aware of and understand the possible risks of HRT. If you are about to start HRT and are worried about side effects and other risks, please discuss it with your physician.

KEY POINTS OF CHAPTER 3

- Menopause has a wide range of symptoms across a wide age range.
- You *may* need a blood test to measure FSH as part of the diagnosis.
- 'Do-it-yourself' menopause diagnosis kits are not recommended.
- Hormone Replacement Therapy (HRT) replaces the hormones which are no longer being produced in the perimenopause/menopause.
- There are many ways of taking HRT, and each has pros and cons.
- There are also many HRT 'treatment routines,' and each has pros and cons.
- All medications have side-effects, and HRT is no exception.

CHAPTER 4

The Alternative Treatments of Menopause

(Some Different Ideas)

Often what feels like a terrible change, is really just a pathway to a far better place.

Karen Salmansohn

INTRODUCTION

The previous chapter explored HRT, which is considered the 'standard' treatment for menopause across the world. Nevertheless, there are options for the treatment of the menopause which are worth exploring and may be very beneficial for some women (but once again, not for all). These alternatives may interest those women who cannot take HRT for medical reasons or simply decide that HRT is not for them. You are allowed to make this decision. No one has to take any medication unless they are 'sectioned' under the Mental Health Act (2007). HRT is not compulsory; it is your choice whether or not to take it following advice and support from your healthcare professional. Some women manage perfectly well without it. At the same time, some women would suffer unbearable symptoms without taking HRT.

LIFESTYLE

The concept about changing the course or occurrence of disease by life-style changes is very popular, effective, and certainly worth further exploration. Lifestyle can mean many different things to many different people. The areas described below are considered possible lifestyle changes which may help to control the symptoms of the menopause. It should be remembered that menopause is not a disease; it is a natural biological process. Despite this, the menopause can produce some nasty symptoms, which may in part respond to lifestyle changes. It is those symptoms that, if present, need controlling by either HRT or some other intervention.

EXERCISE

Perhaps the most obvious lifestyle change is to introduce some sort of regular exercise into your life (if this does not already exist). This does not mean a personal trainer and an expensive gym membership. The exercise involved can be as simple as a regular walk of at least 1 mile. This might be part of your commute, or much nicer, it could be through parkland, woodland, hills or coastal areas. One of the nicest long walks I know is around Grafham water in Cambridgeshire. This walk is 11 miles long, but the water and the scenery make it all worthwhile. Despite this, if your first regular walk is half-a-mile long, then this is a great start on which to build. Regular exercise can reduce hot-flushes and improve sleep (being physically tired at the end of the day is a great way to achieve better sleep). Exercise is also well known to improve the overall mood, such as feelings of anxiousness, irritability or depression, which are very common in the menopause. There are of course many other types of exercise to many and varied to mention here. Weight-bearing exercises, for example, can be very beneficial to maintaining bone density and thus reducing the risk of osteoporosis or weak and brittle bones.

DIET

We are much more diet-conscious today than we used to be, even when I was growing up in the 1960's. The health risks of fatty, salty, processed, and sweet foodstuff are very well described elsewhere, and the benefits of reducing or removing these food items are very clear. Obesity accentuates the symptoms and seriousness of every single disease known, and whilst menopause is not a disease, obesity will accentuate the symptoms. If you are unsure how to improve your diet, many websites and books are available. I would suggest using 'medically based' information, such as that found in the NHS in the UK or the Office of Disease Prevention and Health Promotion in the USA.

BE COOL

One of the most distressing symptoms of the menopause is 'hot flushes'. These often happen at night (often called night sweats), so one way to minimize hot flushes is to wear loose and light nightclothes (or none at all!) and ensure that your bedroom is cool and well-ventilated. The recommended temperature for the bedroom is 15.6 to 19.4° C (60 to 67° F). Please also bear in mind that a winter duvet in the summer months will make matters worse. Please ensure you switch to a summer duvet when the ambient temperature outside rises. Those homes with air conditioning (rare in the UK but common in N. America) will also help control night-time temperature.

FOOD

The importance of a good diet has already been mentioned, but in the context of hot flushes, it is also recommended to reduce or remove caffeine from your diet (decaffeinated tea and coffee are readily available in most countries). Alcohol should also be avoided or taken in moderation (1 glass of wine, or equivalent, per day maximum). Alcohol has many unwelcome physiological effects on your body and increases and trigger hot flushes. It is also best to avoid spicy food if possible, as this has also been linked to triggering hot flushes.

STRESS

Menopause is a source of stress which results in severe symptoms in some women, which comes along with mood swings that can be unpredictable and damaging to relationships. There is no 'magic bullet' for this stress, but it is well known that good sleep and regular exercise, as described above, can be very beneficial for some women. Other activities which involve focussed relaxation, such as Tai Chi or yoga, may be very beneficial in reducing stress. There are, of course, pharmacological options for the treatment of stress, but if the problem can be handled by other means this is much better in the long term. It is also important that your partner, family friends, and friends understand the stressful impact of the menopause so that they can support you in the correct way.

SMOKING AND VAPING

Every healthcare professional will advocate stopping smoking for all patients. The smoking habit in all healthcare professionals stands at around 21%, which is a little ironic but perhaps reflects the stress they are exposed to. Vaping is also not recommended. Nicotine intake is the same as cigarettes, and other side effects are also likely when vaping. Giving up smoking and vaping has been shown to help to reduce hot flushes and of course, the other benefits, such as a reduction in the risk of heart disease, cancer and stroke, are an added bonus. If you need help to stop smoking and vaping, then there is a lot of information on the internet, and of course, your GP or Family Physician will be very happy to help.

VAGINAL DRYNESS

This distressing symptom of the menopause has already been mentioned earlier, but it also fits into the life-style changes which may help the symptoms of the menopause. It is possible to buy vaginal lubricants and moisturisers from most pharmacies (without a prescription), and these products can be extremely effective. It may take a little time to get used to using such products during sexual intercourse, but they will certainly make the whole process much more enjoyable.

PHARMACEUTICALS

There are other pharmaceuticals available, most often by prescription only, which may help those women who find standard HRT either unwelcome or unpleasant. All of these products must be taken under medical supervision, and it is also important to note that some of these pharmaceuticals are not licensed for the treatment of menopause. This means that they have not been through a clinical trial on menopausal patients, and the safety and efficacy in this context have, therefore, not been proven.

TIBOLONE

Tibolone is used in the treatment of osteoporosis and menopause, containing both oestrogen and progestogen. It comes in tablet form and is taken once a day. Some studies have shown that those patients taking Tibolone are at a greater risk of stroke. Tibolone may help in the reduction of hot flushes, low libido and low mood, although some studies suggest that it is not as effective as standard HRT. In addition, Tibolone should only be prescribed to those women who had their last period more than a year ago, *i.e.*, post-menopausal women. As with all medications, there are side effects which include breast tenderness, itching, vaginal discharge and abdominal pain. If you think that Tibolone may be helpful to you, then please discuss it with your GP or family physician.

ANTIDEPRESSANTS

It is no surprise that antidepressants are a possible pharmaceutical intervention in menopause because one of the possible symptoms of menopause is a low mood or even depression. There are two main types of antidepressants relevant to menopause, and these are:

1. Selective Serotonin Reuptake Inhibitors (SSRI)

2. Serotonin-Noradrenaline Reuptake Inhibitors (SNRI)

These medications are licensed for the treatment of depression but not for the treatment of menopause. Nevertheless, some patients report that their hot flushes decline in occurrence and intensity when taking these medications. There are, of course, possible side effects which include reduced libido, nausea, anxiousness and dizziness. More research is needed on the use of antidepressants in the treatment of the symptoms of menopause other than depression, but if you feel that these medications might be of help to you then please discuss it with your GP or Family Physician.

CLONIDINE

Clonidine is a medicine available by prescription only, and it is used to treat high blood pressure. It is a tablet and is taken 2 or 3 times every day. Clonidine does not change hormone levels, but there are some reports that it might have very small benefits in terms of reduction of the symptoms of menopause, especially hot flushes and night sweats. There are side effects, including depression, constipation, dry mouth and general tiredness, and it may take 2-4 weeks before any benefits are seen by menopausal women. Once again, if you think that Clonidine may help you, then please discuss it with your GP or family physician.

BIOIDENTICAL OR 'NATURAL' HORMONES

These are hormones made from plant sources that some people consider similar or even identical to human hormones. There are claims that these products are 'natural' and safer than standard HRT. Despite this, there are several problems with bioidentical hormones, which are:

- They are not currently regulated, and it is, therefore, unclear about the safety and efficacy of bioidentical hormones.
- There is no clear evidence that bioidentical hormones are any safer than standard HRT
- There are no studies available on the effectiveness of bioidentical hormones in the treatment of menopausal symptoms
- The hormones in bioidentical hormone preparations are based on the hormone levels in saliva. There is no evidence to show that this has any relevance at all in the treatment of menopausal symptoms.

What is clear is that bioidentical hormones are not the same as the naturally occurring hormones in the body. In general terms, for the reasons above, bioidentical hormones are not currently recommended for the treatment of menopausal symptoms. Until evidence is produced to the contrary, you are well advised to avoid them.

COMPLEMENTARY THERAPIES

Many complementary therapies are available for a wide range of diseases and conditions. The problem is that few, if any, have been properly tested for safety and efficacy and, therefore, may be just providing some benefits by the placebo effect. The placebo effect is when a beneficial effect is produced by a placebo (an inactive substance) or an unproven treatment which is due to the patient's belief in the inactive substance or treatment. A simple example would be giving a sugar tablet to treat a headache but not telling the patient that it is a sugar tablet. If the patient reports a benefit from the sugar tablet (*e.g.,* the headache goes away), this is the placebo effect. It is, therefore, the patient's belief that the treatment will work, which gives the benefit, not the treatment itself. No one really understands the true mechanism of the placebo effect, but it has been demonstrated many times. It could be, for example, that natural endorphins (pain killers) are released when the patient thinks that he or she has received treatment and therefore, a benefit is felt. The placebo effect must be considered when taking any preparation or product that has not been scientifically proven.

HERBAL REMEDIES

There are many herbal remedies available that are claimed to be beneficial in the treatment of menopause. These include St John's wort, evening primrose oil, ginseng, black cohosh, angelica, and many more. If you want to explore these options, it is probably best to see a herbalist and get their recommendation. It is possible to go into a health shop and simply buy any of these products off the shelf, but it may be a little overwhelming with what is on offer, and the shop staff will not necessarily have the experience or training to help you in your choice. Some reports have reported that herbal remedies such as St John's wort and black cohosh may reduce hot flushes, but the science behind these claims is rather shaky. The problems are that the doses involved are usually very vague, and no one knows if the reported health benefits last. Another serious note of caution is that St John's wort can interact very badly and even dangerously with prescribed medication such as:

- SSRIs and monoamine oxidase inhibitors (MAOIs), including Phenelzine or tranylcypromine
- Tricyclic antidepressants, including Amitriptyline, nortryptyline, or imipramine
- Escitalopram
- Citalopram
- Fluvoxamine paroxetine, fluoxetine and sertraline

The same applies to anti-histamine medication (taken for allergies and hay fever), digoxin (taken for heart disease), and immune-suppressant medication (given, for example, to transplant patients). In summary, if you intend to try any herbal remedy to treat the symptoms of menopause, it is essential to speak to your physician first so that you can avoid dangerously, or even possibly fatal, interactions. Having said that, if you get clearance to take an herbal remedy and it gives you a good benefit, then that is fantastic. Everyone is different; proceed with caution, and if it works, then good for you!

ACUPUNCTURE IN THE TREATMENT OF MENOPAUSE

There have been a few reports that acupuncture may be able to moderate or even remove some of the symptoms of the menopause. A recent study claimed that acupuncture may reduce night sweats, hot flushes, sleep disruption and mood swings. The problem with this, and many other studies on acupuncture for various diseases and conditions, is that in all of these studies, the placebo effect could not be ruled out. Supporters of using acupuncture in the treatment of menopausal symptoms often claim that acupuncture is a 'safe, cost-effective and simple procedure with very few side-effects'. Some acupuncturists claim that when acupuncture needles are inserted, parts of the brain 'light-up', thus changing the way in which the brain responds to the menopause. In short, the evidence about the physiological action of acupuncture is, at best weak, and I would advise caution on relying on acupuncture as a reliable treatment of the menopause.

Hot Flush and Night Sweat Treatment using Acupuncture

Acupuncture has a long, complex, and often confusing history and is part of what is known as 'Traditional Chinese Medicine'. The foundation of these beliefs is that yin and yang (two opposite forces) must be in balance in order to remain healthy. So far, so good, except what exactly is yin and yang? Yin and yang are certainly literally and figuratively far from Western thinking, but this does not necessarily mean that they have no merit. Perhaps just the opposite. The next concept is that when an acupuncturist assesses a menopausal patient for hot flushes and night sweats, the relationship of yin and yang in her kidneys is of paramount importance. This is where problems arise between Western medicine and Chinese medicine because the kidneys are not involved in temperature regulation in the body. The kidney functions are to control acid/base balance in the blood, water and electrolyte balance, remove toxins and waste products from the blood, control blood pressure, produce hormones (not those related to menopause) and activate vitamin D. It requires a great leap of faith to believe that such an approach by acupuncture may intervene in either the severity or occurrence of night sweats and hot flushes. Despite all of this, if you decide to try acupuncture in this context and it works for you, then carry on. There are many things we do not understand in medicine, and if a benefit is found in acupuncture, then there is no reason to reject it. No harm will be done, and your symptoms may decline.

Pain Treatment Using Acupuncture

The menopausal patient may often suffer from pain in specific areas, *e.g.,* the back, or pain in many different areas across her body. The American College of Physicians has studied acupuncture in the treatment of lower back pain and has recommended it to patients suffering from lower back pain as a form of pain control. This is not just menopausal women; it is all back pain patients. If you suffer from lower back pain, or indeed pain in any area of your body, then acupuncture is certainly worth considering.

Mood Swings and Anxiety Treatment Using Acupuncture

There has been a research that shows acupuncture alters neurotransmitters (chemical messengers) in the brain. These changes may reduce levels of anxiety and nervousness in the patient and therefore improve overall mood. Many patients have claimed that acupuncture gives them a sense of peace and relaxation with muscle relaxation and no racing thoughts in their minds. Once again, it is difficult to rationalise these responses to acupuncture, but if you find a benefit, please go ahead and enjoy it. There will be no harm done and no known side effects.

Insomnia Treatment Using Acupuncture

There have been clinically significant reports that cognitive behavioural therapy (CBT) and acupuncture may benefit patients suffering from insomnia. Cognitive behavioural therapy (CBT) is commonly known as 'talking therapy' and aims to resolve problems and their associated symptoms by changing the way the patient thinks and behaves. The most common conditions where CBT can be used as part of the treatment process are anxiety and depression, but it may also be helpful in other physical and mental health problems. In terms of the treatment of insomnia, CBT was found to be the most effective therapy, although acupuncture did show statistically significant results. Once again, if acupuncture gives you any relief from insomnia, then there is no reason at all not to use it. It is safe and has no side effects. It is much better than getting into the area of prescribed sedatives for insomnia which are often very addictive, and the dose required to induce sleep increases with time.

Vaginal Dryness Treatment Using Acupuncture

Some acupuncturists claim that vaginal dryness may be treated using acupuncture to stimulate neurochemical release, which may counteract the imbalance present. There is no scientific evidence at all to support this claim. Nevertheless, if you choose to try acupuncture as a treatment of vaginal dryness and it works for you, then this is a great result!

PLANT BASED EXOSOMES

Exosomes are tiny 'packets' of information naturally released by both plant and animal cells for cell-to-cell communication and regulation. It has recently been shown that the information carried by both plant and animal exosomes is the same, and therefore, plant-based exosomes can be used to regulate animal and human cells. It is important to remember that approximately 40% of all pharmaceuticals currently in use are plant-based, so whilst this concept initially sounds a little weird, the use of plant-derived treatments has been in routine use for many years.

The use of plant-based exosomes as a 'cosmeceutical' has enormous potential benefits. There are two main reasons for this:

1. Studies are already showing benefits in the use of plant-based exosomes to assist in hair re-growth, tissue repair, burn repair, and as a potential anti-aging technology.

2. Plant-based exosomes are supplied in oil which is applied directly to the skin and absorbed through the skin. Future delivery systems may include creams or even patches. This means that plant-exosome therapy is totally non-invasive and natural.

The use of plant-based exosomes clearly has enormous potential to help with the symptoms of menopause. Plant exosomes, which are absorbed through the skin, can travel throughout the body, including the ovaries, testicles and brain. Further studies are needed, but this unique technology will no doubt be in routine use in years to come.

CONCLUSION

This chapter has explored various 'complementary' and 'alternative' approaches to the treatment of the symptoms of menopause. New ideas and concepts are always arising, and it is important to keep up-to-date with these changes. If you try any non-standard treatment that benefits you, then this is great for you. It may not necessarily have the same benefit in other people, but if you benefit, then there is no reason not to use 'complementary' and alternative 'therapies' to manage the symptoms of your menopause.

KEY POINTS OF CHAPTER 4

- Some lifestyle changes may help to alleviate symptoms of the menopause.
- Prescribed alternative medication is available to help to treat the menopause.
- Bioidentical or 'natural' hormones may be helpful to some women.
- Complementary therapies may be useful, but they should always be used under medical supervision.
- Plant-derived exosomes may be very important in the future management of menopausal symptoms.

CHAPTER 5

The Male Menopause

(Fact or Fiction?)

Acting feels different. I'm not sure exactly what that is, but it used to mean a lot more. Maybe that sounds like I'm throwing it away, and I'm not, I'll still do the best damn job I can, but it doesn't mean the same thing. I'm going to get the answer for myself one of these days. It's the male menopause, that's what it is.

Mel Gibson

INTRODUCTION

There is no such thing as the male menopause. You may feel from this statement that this is going to be a very short chapter. If the male menopause is to exist, then it has to be called the male andropause. This at least makes sense of the terminology. Many things happen in the male body as it ages, but there is no male analogy to female menopause. Despite this, men do undergo physiological and biochemical changes when ageing, which may explain why some may feel that they might be suffering from 'menopause'.

MALE AND FEMALE BIOLOGY

The reason why the 'male menopause' does not exist can be found in basic biology. The female is born with a finite number of eggs. From puberty, some of these eggs are recruited every month to mature to the point where they may be fertilised. If fertilisation does not take place, then these eggs die. This means that every month the reserve of eggs a woman holds slowly decreases, and eventually (at menopause), no more eggs are available, and the ovaries stop functioning. This process results in the menopause symptoms described earlier in this book.

The reproductive mechanisms of the male are very different biologically from the female. The male produces sperm in the testes from puberty. These sperm are created from special stem cells called spermatogonia. This process of sperm production takes about 64 days, and because it is driven by self-replicating stem cells, the testicles never run out of sperm and function for the whole life of a man. This is why a man can father children in old age whilst a woman cannot become pregnant in old age. The biological changes which a woman undergoes are therefore not replicated in any way in a man making the term 'male menopause' terminology which does not make any sense.

THE MALE ANDROPAUSE

Male andropause is a condition that is becoming mainstream in current medical thinking, and it all relates to the hormone testosterone. Testosterone in a man is a hormone responsible for muscle mass, facial and body hair, and of course, the lower voice of the male. The production of testosterone from the testicles (under control by the brain) starts at puberty, and the physical effects can be seen very easily as a boy develops into a man. Despite this, as men get older, the levels of testosterone in the blood slowly decrease, which sometimes results in fewer sperm being produced (but sperm production does not stop) and some physical and psychological changes. Testosterone levels decrease at about 10% every decade from the age of 30, and this is part of the natural male ageing process.

The key point about male andropause is that the decrease in testosterone is gradual, whereas the decline in hormones in female menopause is rapid. It is estimated that 30% of men in their 50's may develop symptoms of the andropause.

MECHANISM OF THE MALE ANDROPAUSE

To understand the mechanism or mode of action of the male andropause requires more understanding of male hormone interactions. It has already been established that the male hormone, testosterone, is constantly declining from about the age of 30. In parallel to this, another hormone called sex hormone bonding globulin (SHBG) begins to rise. The action of SHBG binds to testosterone in the blood, and bound testosterone is inactive. This means that less biologically active testosterone, known as bioavailable (or free) testosterone, is available to use in the body. The overall effect is that testosterone levels fall. It has been clearly shown that men who experience the symptoms related to andropause have low levels of bioavailable testosterone in their blood. The tissues in the body which rely on testosterone for their normal activity receive less and less testosterone, and it is this which is proposed to cause the physical and possibly mental changes in an andropausal man.

THE 'MID-LIFE CRISIS'

Some men undergo what has become known as the 'mid-life crisis'. The main symptoms of this seem to be an irresistible urge to buy a two-seater sports car or, worse, a high-power motorbike. There is also an increased chance of such a man seeking an affair with a younger woman. No one understands this behaviour, least of all the men involved, but it may be part of the andropause, which is influencing changes in the brain and the resultant poor decision-making. Official statistics in the UK show that male motorbike riders aged between 41 and 55, riding a motorbike with an engine size of 500cc and over account for 41% of fatalities and serious injuries. This spike in motorbike fatalities and serious injuries coincides with the 'mid-life crisis'. Whether or not the 'mid-life crisis' exists, and if it has some form of physiological basis making it part of the andropause is unknown. More research is needed but in the meantime, steer clear of grey-haired bikers riding powerful motorbikes with a young woman on the back!

SYMPTOMS AND COMPLICATIONS OF THE ANDROPAUSE

As with the menopause, the symptoms of the andropause will vary greatly from person to person. In general, the symptoms are all related to the ongoing decrease in testosterone and include the following:

- General tiredness
- Low libido
- Erectile problems
- Mood swings
- Depression
- Loss of muscle mass
- Body fat increase
- Hot flushes

These symptoms may look familiar. Some of them, such as erectile dysfunction and a loss of muscle mass, are clearly male orientated but the rest could easily be found on a list of the symptoms of the menopause. This is because both andropause and menopause result from a significant change in the levels of key hormones (known as the sex hormones). The similarities between the symptoms of the menopause and andropause are reflected in this underlying cause. The complications which are associated with the andropause include an increased risk of cardiovascular disease and osteoporosis, which are once again similar to the menopause.

DIAGNOSIS OF THE ANDROPAUSE

The primary problem with the diagnosis of the andropause is the reluctance of older men to visit their GP or Family Physician. This raises problems not only for the diagnosis of the andropause but for other unrelated diseases which may be either life-changing or life-threatening. The reasons for this reluctance of older males to consult with primary healthcare professionals are unknown, but there are possible underlying psychological reasons. Perhaps the older male fears the possible diagnosis or treatment which may result from seeing a healthcare professional. Perhaps they prefer to 'suffer in silence'? Spouses and partners of these men may argue otherwise! Perhaps the male psyche of 'I am a strong man, and I don't get diseases' plays a role? The result of this reluctance by older males to see primary healthcare professionals often results in the first interaction of the patient with the healthcare system by attending the Accident and Emergency Department of a hospital when their symptoms become intolerable. The basic message here is that if you feel unwell (for any reason), please see your GP or Family Physician. It might save your life.

Getting back to the andropause, if the male patient plucks up the courage to see a primary healthcare worker, then specific questions will be asked and an examination will be carried out, relating to the symptoms of low testosterone levels. The confirmation of the diagnosis is by a blood test to measure the levels of testosterone in the patient. The expected normal amounts of free or biologically active testosterone should be:

- Age 45-54: 91 pg/mL
- Age 55-64: 83 pg/mL
- Age 65-74: 69 pg/mL

A pg (picogram) is one trillionth (10^{-12}) of a gram and mL are millilitres. The range for the bound (or biologically inactive) testosterone is higher, although the decline with age is the same. Any patient who has a free testosterone level below the normal levels given above may suffer the symptoms of andropause, but equally, many patients with low levels do not have symptoms (or perhaps ignore the symptoms!).

There are other diseases that are associated with low testosterone levels, such as diabetes and high blood pressure (hypertension), which will have to be excluded before a diagnosis of andropause can be made. It must also be known that the symptoms of andropause are often seen in the natural ageing process, so treatment may be ineffective in these cases.

TREATMENT OF THE ANDROPAUSE

As in menopause, the treatment for the andropause is to replace the missing hormone(s), which in this case is testosterone. This may improve the quality of life and reduce the symptoms. In parallel with the menopause, other interventions such as increased exercise, a healthy diet and stress reduction activities such as yoga may also help. As with the menopause, there are many treatment routes available to replace the missing testosterone, these include:

- Skin patches. These work in the same way as the skin patches used to treat the menopause, but in this case, they contain testosterone. This route allows a slow and steady release of testosterone into the bloodstream, and most of these products require a new patch to be applied every day.
- Capsules. These are taken twice daily after meals. These capsules are unsuitable for men suffering from heart or kidney disease, liver disease or high levels of calcium in the blood.
- Testosterone Gel. This is once again very similar to the gel treatment of menopause, but the main constituent of the gel in andropause is testosterone.
- Testosterone Injections. This is an injection of testosterone into muscle every 2-4 weeks. These injections may cause mood swings but provide a regular and steady dose of testosterone and may be suitable for some patients.

Your physician will tell you that men suffering from prostate or breast cancer (yes, men get breast cancer) must not take testosterone. In addition, those men suffering from heart disease, who take anti-coagulants (blood thinners), have liver or kidney disease, have an enlarged prostate, or have too much calcium in the blood must not take testosterone.

Andropause is clearly a medical reality, but it does not apply to all men, and the symptoms may be very variable or even non-existent. More research is needed to fully understand andropause and to develop novel treatments in those men who suffer symptoms.

KEY POINTS OF CHAPTER 5

- The 'male menopause', if it exists, must be referred to as the andropause.
- There are some possible hormonal changes that could result in the andropause.
- The symptoms of the andropause are complex and vary from man to man.
- The levels of biologically active testosterone may correlate with the andropause.
- The andropause may be treated by testosterone therapy, but this must be supervised by a healthcare professional.

CHAPTER 6

Regenerative Medicine and the Menopause

(The potential use of stem cell technology to alleviate the menopause)

The regenerative medicine revolution is upon us. Like iron and steel to the industrial revolution, like the microchip to the tech revolution. Stem cells will be the driving force of this next revolution.

Cade Hildreth

INTRODUCTION

Regenerative Medicine is now recognised as a branch of medical research and practice, and in fact, I am a Professor of Regenerative Medicine based in Cambridge, UK. The process of Regenerative Medicine involves the use of stem cells (to be explained below) to repair or regenerate tissues within the body. The process of Regenerative Medicine is, in fact, going on naturally every day in the bodies of all of us. For example, stem cells in the bone marrow are constantly replacing dead or 'worn-out' blood cells to keep our blood healthy. In addition, stem cells in the skin are constantly repairing dead or 'worn-out' skin cells. The process of regenerative medicine can be seen most dramatically when the skin repairs itself following a traumatic or surgical injury. Sometimes a scar is left behind where the injury or surgery was, but other than that, the skin has been perfectly repaired (regenerated) by stem cells. I did promise to explain what stem cells are, so here goes: Stem cells are specialised cells found throughout the body of everyone, which when they divide into 2 (which all cells do many times during their lifetime), one of those cells is a 'new' stem cell identical to the mother cell and the other cell is a cell which can go on and repair or replace damaged or old tissue. The important thing about this statement is that the stem cells, when they divide, produce another stem cell; in scientific terms, this is called self-replication. Self-replication means that stem cells can continue to repair the tissues in our bodies for our whole life span. It is true that these stem cells in our bodies will age as the rest of the body ages and may be less capable of repairing tissue in 'old age'. This could be the basis of ageing in humans which, once again, we can see only too clearly in the skin as a person gets older. This quick summary of Regenerative Medicine gives the basic points, but if you wish to learn more about Regenerative Medicine, then you might like to read my book 'The Regeneration Promise', which is in the suggested reading list at the end of this book.

STEM CELLS AND TREATMENT OF THE MENOPAUSE

At present, possibly the most researched stem cells in terms of using them as a potential treatment for the menopause are known as Mesenchymal Stem Cells (MSC). These MSC are found in many tissues of the body, including fat tissue, bone marrow and the umbilical cord, which provides a link between the baby and the mother during pregnancy. They are even to be found inside teeth! MSC are interesting in that they are known to be able to repair certain tissue tissues such as bone, fat, and connective tissue (*e.g.*, tendons), and they may be able to repair nerve tissue. MSC have been used the most to treat inflammatory diseases such as arthritis and to repair damaged joints. The MSC technology in this context does seem safe and effective. In this particular example, the MSC are injected directly into the joint. A useful advantage of MSC is that they can be donated from one person for use in another person without the risk of rejection; this is because the MSC does not carry the molecules on their surface (Class II HLA), which induces rejection. The MSC can also be collected, processed and frozen for later use if required. The problems with MSC as a source of stem cells for the treatment of menopause are:

- The collection of MSC for use is invasive for either the patient or the donor. For example, fat is obtained by abdominal liposuction, which whilst not too unpleasant (a needle is introduced into the abdomen and the fat is sucked out), is an invasive procedure with all the usual risks of infection and bleeding. Bone marrow collection (by introducing a very sharp surgical steel needle into the pelvic bone) is even more invasive and unpleasant.
- The processing of MSC is different in different laboratories, and as such, the process cannot be standardised for clinical use. When any treatment is used clinically, then the production of that treatment must be standardised, reliable and reproducible. The is to ensure the safety of the patients receiving the treatment.
- MSCs are derived from a wide range of sources, and each MSC type has slightly different properties, so identifying the 'best' MSC to treat menopause will be a long and expensive process.
- The laboratory technology, expertise and regulatory licensing required to use MSC clinically are very expensive.
- The way in which the MSCs are given to the patient is invasive (often by an injection directly into the ovaries), and the dose required (*i.e.*, the number of

stem cells needed for effective treatment) is currently left to either availability or guesswork!

In summary, MSC has some potential in the future treatment of the menopause, but there is still a lot of time-consuming and expensive work to do if they are to become a standard treatment for the menopause. The worry is that this time and investment may lead nowhere in the treatment of the menopause because of cost and technical difficulties.

Very Small Embryonic-like (VSEL) Stem Cells In The Treatment Of The Menopause

Please try to stick with the section because it contains information that may have an impact on every menopausal woman in the world. It is arguably the most important section of this book. What I will discuss here is potentially life-changing for billions of women. It has some science along the way, but I will explain it as we go, so please do not feel daunted to continue reading. The science is actually easy, and the implications are enormous.

Before I proceed with a description of this concept (which is based on a long collaboration between myself and my dear colleague and friend Dr. Todd Ovokaitys in Carlsbad, California, USA), there are two extremely important points to make:

1. VSEL stem cells are very small. They are, in fact, around 2-4 μM (microns) in diameter (a micron is one-millionth of a metre). This means that they can easily cross the blood-brain barrier, which is a biological barrier protecting the brain from the rest of the body. The result of this is that VSEL stem cells can very easily pass from the blood into the brain and can therefore be active within the brain. All other stem cells are too large to do this, and if needed in the brain, then they have to be directly injected into the brain (not much fun for everyone involved). In addition, VSEL stem cells can cross the blood-testes barrier, a similar biological barrier protecting the testes from the rest of the body. Once again, this means that VSEL stem cells can easily enter the testes and carry out the repair if required. VSEL stem cells can also cross the blood-follicle barrier, which is a biological barrier in the ovaries. This means that the VSEL stem cells can easily get into every biological part of the ovary and carry out the repair if needed. These facts will help to understand the proposed mechanism of action of VSEL stem cells which I will describe below.

2. VSEL stem cells are 'embryonic-like'. This does **not** mean that VSEL stem cells are embryonic stem cells. They are simply 'like' embryonic stem cells because they carry some of the same molecules on their surface as embryonic

stem cells. Embryonic stem cells are derived from human embryos and are not connected in any other way to VSEL stem cells other than sharing some similar cell surface molecules.

VSEL Stem Cells in the Human Body

Most scientists agree that VSEL stem cells were remnants of our embryonic stage of development when we were all about 1-2 mm in diameter! From a biological viewpoint, VSEL stem cells are often proposed to be the 'stem cell of stem cells'. This means that they may be the stem cells that created the whole range of stem cells we see in a fully developed human that keeps us all alive and well daily. VSEL stem cells have been found throughout all of the tissues in the human body, including in the circulating blood in our veins and arteries. This makes VSEL unique amongst all other stem cell types in that they can easily be obtained from the circulating blood, *i.e.*, from blood in a vein. This means that a very simple blood collection (which just about every human being on Earth has experienced at some time or another) is an easy source of perhaps the most important and least researched stem cells we know about at present. The only other stem cell which can be collected from the circulating blood are known as peripheral blood stem cells (PBSC), which are 'mobilised' from the bone marrow by strong medications. These 'mobilised' bone marrow stem cells can then be collected from the circulating blood using sophisticated and expensive technology called apheresis and can be used to treat blood disorders such as leukaemia. This is a great way to treat leukaemia, but it is not relevant to the treatment of the menopause. VSEL stem cells can be collected using no medication, no expensive equipment, and just a syringe and needle. This makes VSEL stem cells the most accessible, cost-effective stem cells in the human body. This is a very important point. In general, stem cell technology is highly complex and very expensive. VSEL stem cell technology is easy to understand and relatively cheap to use.

Pluripotent VSEL Stem Cells

Once again, please try to stick with this brief amount of science because it is critical in understanding the considerable potential of VSEL stem cells. VSEL stem cells are 'pluripotent'. This means that VSEL stem cells are capable of producing every tissue in the body. This means that, in theory, VSEL stem cells *can repair any type of tissue in the body*. If the patient has a diseased heart, then VSEL stem cells can enter the heart from the circulating blood and repair the damage to the heart. If the patient has a neurodegenerative disease, such as Alzheimer's disease, then VSEL can enter the brain from the circulating blood and repair the nerve cells damaged by Alzheimer's disease. If a patient has the menopause, then VSEL stem cells can enter both the ovaries and the brain to repair and restore the physiological damage caused by menopause. This might come as a bit of a shock to those women who are reading this and suffering from the symptoms of the menopause or the side effects of HRT. We have no reason to think that this concept is not true, and indeed we already have preliminary data showing that menopausal women benefit from VSEL stem cell treatment in many ways.

VSEL Stem Cells in the Blood of Everyone

I have now established that VSEL stem cells are pluripotent (can make any tissue in the body) and that they can easily be obtained from the circulating blood. Our own research, and that of others, has shown that the VSEL stem cells which are in the circulating blood appear to be inactive or 'dormant' from a biological viewpoint. This means that they do not seem to contribute anything to our daily health and well-being. I have to admit that this is a little confusing, as if something exists, it usually exists for a reason. Other workers have shown that the VSEL stem cells in other tissue within the body (apart from the blood) may be biologically active during daily life. This could, for example, explain the fact that the liver (an organ that carries out billions of chemical reactions essential to life every day) has the ability to regenerate itself under certain conditions. I should say that all of these concepts are very much open to debate. Science can be a very frustrating thing, but when it works, it is amazing.

Activation of VSEL Stem Cells Derived from Circulating Blood

Since the VSEL stem cells in the circulating blood, appear to be 'dormant', and, therefore, unlikely to benefit anyone in terms of a treatment, my colleagues and I have explored ways in which they may be biologically 're-activated' and therefore become useful in treatment. A physicist and colleague in Edinburgh called Dr. Scott Strachan (who has sadly now died) developed a novel way of 'modulating' or changing the physical properties of laser light. This was later further developed into a low-power red laser to use in conjunction with VSEL stem cells, and it is now called the QiLaser (the QiLaser is fully patented). Our subsequent research has shown that the QiLaser can activate 'dormant' VSEL stem cells found in circulating blood by 'opening' otherwise 'closed' molecules on the surface of the VSEL stem cells. These 'activated' VSEL stem cells have some very remarkable properties in terms of the treatment of disease.

How is all of this Relevant to The Menopause?

First of all, thank you for suffering the science above. I hope that it was not too boring, but it may have been slightly confusing. Despite this, it allows me to set out how VSEL stem cells could be used to treat the Menopause. The process would be very simple, following these steps:

- A menopausal woman (or even a woman who is in the perimenopause, or premature menopause (often medically referred to as Premature Ovarian Insufficiency, POI) would be a perfect candidate for the VSEL treatment (known as the QiGen Protocol).
- The patient would have to attend one of the clinics currently offering the QiGen Protocol (currently mostly in N.America, but we aim to be global as soon as possible) and be seen by a physician at that clinic to confirm that she is suitable for treatment using QiLaser activated VSEL stem cells (known as the QiGen Protocol).
- On the day of treatment, the patient attends the clinic for no more than 2-3 hours for the QiGen Protocol.
- The first step is to take 66 mL of circulating blood from the patient by inserting a small, temporary catheter into the arm vein of the patient. This so-called 'intravenous line' remains in place during the QiGen Protocol and is used to take blood and to return QiLaser-activated VSEL stem cells to the patient.
- The 66 mL of blood (in 6 special tubes called Platelet Rich Plasma PRP tubes) is then centrifuged so that the red cells (oxygen-carrying cells in the blood) are separated from the plasma and platelets, which constitutes PRP. Each PRP tube produces about 7 mL of PRP, and a typical preparation therefore results in a total volume of about 42 mL PRP.
- PRP contains high numbers of platelets (they are the fragments of a cell called a megakaryocyte which is found in the bone marrow and not cells in their own right). Platelets are involved in blood clotting but can also produce many other molecules described below, which can be very beneficial.
- PRP also contains growth factors and cytokines (some of these are produced by platelets). These are natural chemicals found in the blood which are involved in the promotion of cell growth and tissue repair.
- Perhaps most importantly, PRP contains many VSEL stem cells (typically around 63 million VSEL stem cells in the total volume of PRP prepared).
- Once the PRP has been prepared, it is exposed to the QiLaser for 3 minutes. Our research has shown this exposure of the QiLaser to the PRP for 3 minutes is the optimal exposure time to maximise the benefits to the patients.
- The QiLaser-activated PRP is then returned (at about 1-2 mL per minute) intravenously into the patient using the same intravenous line which was put in place to collect the blood.
- During infusion of the QiLaser-activated PRP, the physician may also apply the QiLaser to the patients' body in the areas where stem cells are needed. For example, in the case of the menopause, the QiLaser may be applied to the area

of the ovaries and uterus (womb). We have preliminary data indicating that applying the QiLaser to the body may help 'direct' activated VSEL stem cells to those areas in the body where they are needed. Further work is needed on this concept, but in patients treated so far (for a wide range of diseases), applying the QiLaser to the body seems to help. I should mention that the QiLaser is totally harmless and does not have enough power to damage human tissue.

- Once the activated VSEL stem cells have been re-infused, the intravenous line is gently removed, and the patient is monitored for the next 30 minutes to ensure no unexpected problems arise.
- The patient is then free to leave the clinic.

To date, thousands of QiGen Protocols have been carried out without any adverse effects for many diseases ranging from heart disease to neurological disease and even neurological trauma (*e.g.*, spinal cord injury) with impressive results.

Many patients experience very rapid benefits after receiving the QiGen Protocol. We believe (but have yet to prove) that these very rapid benefits, often seen as quickly as within 5 minutes of receiving the Qigen Protocol, are due to the action of the cytokines and growth factors present in concentrated form in the PRP. These very rapid effects happen too quickly to be cell-based, and most are transient benefits. This also fits in with the idea that these rapid effects are growth factor and cytokine based. In contrast, most cell-based benefits in regenerative medicine (*e.g.*, a bone marrow transplant) may take weeks or even months to become apparent and are permanent. Patients who receive the QiGen Protocol often experience benefits in the days and weeks following the QiGen Protocol, which we believe to be based on the activity of QiLaser-activated VSEL stem cells. The benefits tend to be long-term because the QiLaser-activated VSEL stem cells will remain present and active for many months, years, or even a whole lifetime.

How May the QiGen Protocol help in Menopause?

The QiGen Protocol may be of great benefit to menopausal women for several reasons; these include:

- The VSEL stem cells are obtained from the circulating blood of the patient herself, and once activated, are returned to the same patient intravenously. This means there are no complex collection procedures, tissue matching, or infectious disease screening requirements. The patient receives her own Qilaser-activated PRP.
- The QiLaser interacts with stem cells, platelets, growth factors and cytokines in PRP. We have published a scientific paper describing how the QiLaser interacts

with VSEL stem cells and modifies the expression of cell surface molecules (the link to the publication about this is in the Useful Links section at the end of this book).

- We have, in addition, published a mechanism for the mode of action of the QiLaser on VSEL stem cells using theories taken from Quantum Physics. This is the first publication of its' type where Quantum Physics has been used to explain the interaction of modulated laser light with a living cell (the link to the publication about this is in the Useful Links section at the end of this book).

- The QiLaser will also interact with platelets, growth factors and cytokines in the PRP. At the present time, we do not have any publications on these interactions, but it seems reasonable to suggest that these complex molecules will be activated in some way when the QiLaser interacts with them. This is a subject of ongoing research to fully understand how the QiLaser interacts with every component in PRP.

- On re-infusion of QiLaser-activated PRP into the general circulation of a menopausal patient, 3 main events occur:

1. The QiLaser-activated VSEL stem cells enter the general circulation of the patient *via* the intravenous line which was initially used to take blood to make PRP. The VSEL stem cells can easily enter the ovaries from the blood due to their small size, (possibly directed by the QiLaser and by their own natural homing ability), and begin 'repair' of the ovaries. This travelling of stem cells to the place where they are needed is called stem cell homing. Another example of stem cell homing is the homing of bone marrow stem cells from the circulating blood to the bone marrow following a bone marrow transplant. It is a completely accepted concept in the field of cell therapy. In terms of QiLaser-activated VSEL stem cells, we have shown that the QiLaser activates the homing molecules on the cell surface of the VSEL stem cell. Once inside the ovary, the QiLaser-activated VSEL stem cells because they are pluripotent, can repair the structure of the ovary and may even be capable of repopulating the egg reserve within the ovaries. This repair of the ovaries in this way could represent a significant benefit to the menopausal patient.

2. The second important organ which is involved in the overall symptoms of the menopause is the brain. The QiLaser-activated VSEL stem cells can easily enter the brain from the circulating blood by crossing the blood-brain barrier. Once inside the brain, the QiLaser activated VSEL stem cells can repair and replace neuronal tissue that has either been damaged or changed due to the menopause. Such changes and repair within the brain, along with the repair of the ovary, could have considerable benefits to menopausal patients.

3. Finally, the growth factors and cytokines in the PRP, which have interacted with the Qilaser, will enter every area of the body *via* the general circulation. We believe these growth factors and cytokines result in very rapid (within minutes) benefits to some patients. The QiLaser-activated cytokines and growth factors have a limited life-span. Therefore, any rapid benefit seen which is caused by these QiLaser-activated growth factors and cytokines will be transient.

The combination of QiLaser-activated VSEL stem cell-based ovarian repair, brain repair and the stimulation of repair and regeneration by growth factors and cytokines exposed to the QiLaser represents a potential major step forward in the way in which we think about and treat the menopause. At present, our data on this subject are limited, but based on the data we have seen from life-threatening diseases such as end-stage heart failure treated with QiLaser-activated VSEL stem cells, there is great optimism for the technology in the treatment of the menopause.

VSEL STEM CELL AGE

In the proposed treatment of menopause described above, the VSEL stem cells are taken from the patient herself and returned to the patient in what is known as an autologous treatment. We have shown that this is an effective route to treatment. The reason for this success may be that the VSEL stem cells in the circulating blood are biologically dormant. This means that the VSEL stem cells are not dividing on a regular basis and that the biochemistry within the VSEL stem cell itself is either inactive or greatly reduced. A consequence of this is that the dormant VSEL stem cells in the blood will not be aging with the rest of the body. Aging is a complex process involving many different physiological actions, including changes to the DNA in our cells. When our cells divide, these changes (*i.e.*, aging) are generally considered irreversible. Since the VSEL stem cells in the general circulation are dormant, they are not dividing and, therefore, effectively not aging. These VSEL stem cells in the blood circulation of everyone, may have a biological age equalling the biological age we all had when we were born! Just to clarify this, a person who is aged 60 years old now (or indeed of any age) in terms of chronological age may possibly have VSEL stem cells in their blood which are effectively 0 years old. This concept (which has yet to be proven, but research is underway) could mean that the VSEL stem cells can repair both body tissue and create new stem, which is biologically very 'young' compared to the biological age of the patient. This new tissue and new stem cells will behave biologically as newly formed tissue and cells with no age at all!

'YOUNG' DONOR VSEL STEM CELLS

All of the information provided so far has been about using VSEL stem cells taken from the patient to treat the same patient, *i.e.*, an autologous procedure. This makes the whole process very convenient, relatively cheap and easy to carry out. If our idea that VSEL stem cells in the blood are not aging is true, then they can be used as the 'gold-standard' in this type of technology. Even if we discover that the VSEL stem cells do, in fact, age in the blood, but the outcomes of treatment are still good, then there would still be no benefit in using donor VSEL stem cells. Despite all of this, we have ongoing research to assess the properties of 'young' donor VSEL stem cells, which would be obtained either from young, healthy donors (*e.g.*, from volunteers aged 20-25) or from umbilical cord blood which is the blood left in the umbilical cord and placenta after a baby is born. These sources of 'young' VSEL (known as allogeneic VSEL stem cells) may have advantages over those VSEL collected from the patient herself, and if this proves to be true, then the effort to obtain as use 'young' donor VSEL stem cells may be worthwhile.

HURDLES STILL TO CLEAR

These concepts using QiLaser-activated VSEL stem cells, taken from the patient herself, to treat the menopause clearly need to be confirmed by using placebo-controlled clinical trials. These data will confirm the safety and efficacy of the QiGen Protocol, which will enable it to enter routine clinical practice.

If 'young' donor VSEL stem cells prove to be the most effective way to treat the menopause, then this is a more complex process that would require:

- Identifying, recruiting and obtaining 'young' VSEL stem cells from young donors
- All young donors would have to undergo infectious disease screening, ABO Rh (blood group) typing and tissue typing (HLA typing) to match the recipient.
- Storing of the 'young' VSEL stem cells until needed, probably frozen in liquid nitrogen (quite an expensive process).

If umbilical cord blood is proven to be the best source of 'young' VSEL stem cells, then this process would require:

- Identifying pregnant women willing to donate their cord blood following the birth of their baby (many women may wish not to take part)
- Collecting, processing and storing the VSEL stem cells in cord blood (quite an expensive process)
- All cord blood VSEL stem cells would have to undergo infectious disease screening, ABO Rh (blood group) typing and tissue typing (HLA typing) to match the recipient.

CONCLUSION

There is enormous potential in using Regenerative Medicine technologies to treat menopause in the future. The potential already seen with VSEL stem cells may be the future route to new routine stem cell-based treatments. Nevertheless, caution is needed in that we are in the very early phase of research and development in the use of stem cells for the treatment of the menopause. Hard work is also required, but my prediction is that with the correct focus and investment, stem cell therapy for the menopause will become routine in years to come.

KEY POINTS OF CHAPTER 6

- Regenerative Medicine procedures, using stem cell technologies, may be a possible treatment route for both the menopause and the andropause in the future.
- Mesenchymal stem cells (MSC) may have promise in the treatment of the menopause, but they have preparation and treatment issues to be resolved.
- Very Small Embryonic Like (VSEL) stem cells may have promise in the treatment of the menopause. There is, however, still much research to be done before such technology enters mainstream treatment.
- 'Young' donor VSEL stem cells may have future potential, but once again, much more research is needed.

CHAPTER 7

The Psychology of Premature Menopause, Perimenopause and Menopause

(Mind over Matter)

A woman must wait for her ovaries to die before she can get her rightful personality back. Post-menstrual is the same as pre-menstrual; I am once again what I was before the age of twelve: a female human being who knows that a month has thirty days, not twenty-five, and who can spend every one of them free of the shackles of that defect of body and mind known as femininity.

Florence King

INTRODUCTION

Menopause is a natural process that all women will experience at some stage in their life. The menopause produces many unwelcome physical symptoms in most women, and at present, we manage these physical symptoms by using Hormone Replacement Therapy (HRT). In the future, we might even have safe and effective stem cell-based treatments for the treatment of menopause some of which might even be capable of 'reversing' the menopause. There may even be plant-based 'cosmeceuticals' that could help the symptoms of the menopause. If this comes to fruition, then the management of the physical symptoms of the menopause will change beyond all recognition. Whatever therapeutic approach we seek to follow, at present, the focus is on removing or reducing the physical symptoms of the menopause to enable the menopausal woman to enjoy a better quality of life. This is an admirable goal which I fully support. Despite this, women who go through the menopause not only suffer physical symptoms, but some also suffer psychological symptoms. These psychological symptoms may be severe enough to damage the long-term mental health of the menopausal woman. This subject of the psychology of menopause will be explored in this Chapter to try to better understand this critical component of the menopause, and where possible, to offer help and advice to menopausal patients suffering psychological symptoms.

'THE LUCKY ONES'

In any disease, or biological state, there are always 'the lucky ones'. These are the people who, when diagnosed with a serious disease, undergo treatment, respond brilliantly, get well very quickly, and rapidly return to their former quality of life. Such patients sometimes develop a brighter or more positive outlook on life and may even get involved in fundraising for research into the disease from which they suffered. A classic example of this is cancer. Some cancer patients respond very well to treatment and seem to 'glide' through the whole treatment process without any undue stress. For other cancer patients, the experience is sadly less pleasant.

The same principles apply to the menopause. Some women 'glide' through the menopause, do not require any treatment at all, and continue to enjoy an excellent quality of life with no symptoms. From a scientific viewpoint, we do not have the slightest clue how these women enjoy very easy menopause. Despite this, I am exceptionally pleased with each and every one of them. Future research may help us to explain this 'easy' menopause experience which may even benefit other menopausal patients with more severe symptoms. They are truly the 'lucky ones' in the menopause world. We need much further and deeper research into why some people are 'the lucky ones' when it comes to disease and the menopause. The answer is likely to have a genetic basis, but there will be many other reasons we have not even thought about.

The next category is those women who develop menopausal symptoms but, by taking HRT or using alternative therapies, manage to remove or at least reduce the physical symptoms and therefore return to a good quality of life. They are perhaps 'the fairly lucky ones'. This group seems to be the majority of menopausal women. The final category of menopausal women who suffer severe physical and psychological symptoms, and do not respond well to HRT or alternative therapies. The situation becomes so severe that the mental health of these women also begins to deteriorate, and these women are certainly not 'the lucky ones'.

The Psychology of Premature Menopause or Premature Ovarian Insufficiency (POI)

The diagnosis of premature menopause or premature ovarian insufficiency is a life-changing diagnosis for a young woman. One of the most common causes of premature menopause is the removal of the ovaries and Fallopian tubes, usually because of cancer. Whilst this is a life-saving procedure, such women (once the cancer is clear) may suffer adverse long-term effects, including coronary artery disease, Parkinson's disease, mood disorders, sexual dysfunction and osteoporosis. If any of these complications do occur (and they are not necessarily going to happen to everyone), they also accompany psychological stress.

The second type of menopause in this section is premature ovarian insufficiency (POI), where ovarian function is lost before the natural age of menopause (40-60 years old). Approximately 1% will of women will enter the menopause before the age of 40. The reasons for this may be immune problems, genetic disorders, surgery, and chemotherapy or radiation therapy. In some women, the cause is unknown. From a psychological viewpoint, perhaps the unknown causes patients to suffer the worst as they will constantly wonder if it was 'something they did' that caused premature menopause. This form of 'self-blame' needs urgent support from psychological counsellors, and once again, help can be found initially at your GP or Family Physician.

Sadly, the time to diagnose premature menopause can be very long. The COVID-19 pandemic has exacerbated this, but even before that event, patients often waited months, or even years, before they understood what was happening to them. This is, of course, unacceptable because this long diagnosis time means that advice (such as psychological support) and treatment (such as HRT) are seriously delayed resulting in even more psychological stress for the patient. There is no 'instant cure' for this problem, and there might not even be a long-term cure. Unfortunately, the onus is currently on the patient to drive forward the diagnostic and therapeutic process, which is a sad reflection on the current healthcare provision for premature menopause in most countries.

Mental Health

The menopause, when it occurs at the expected age, can bring many mental health challenges. The premature menopause, which can sometimes arise in women as young as 20, brings even greater challenges to mental health. The primary worry and anxiety is that such a young woman may probably not be able to have children in the future. I say 'may probably not' because the rapidly improving technology in *in vitro* fertilisation or 'test-tube babies' may in the future enable women who suffer premature menopause to still have children. Other psychological aspects, which are particularly common and unwelcome in premature menopause, include difficulty in coping with the altered self-image, the potential loss of fertility and sometimes sexual dysfunction. These are all life-changing events, and the psychological impact of these on the patient must not be underestimated. Advice and support can be found from a psychologist or sex therapist, and once again, your GP or Family Physician will be able to help with such referrals. Organisations such as the Daisy Network provide excellent information, support and resources for women undergoing POI.

The Psychology of the Perimenopause and Menopause

The cause and the physical symptoms of the perimenopause and menopause have been described earlier in this book. I would like to spend a little time exploring the psychological side of the perimenopause and menopause. When the first signs of the perimenopause begin to appear, they are often very mentally stressful for the patient as well as being physically uncomfortable. Typical worries which arise and persist at this time are the fear of the symptoms being responsible for a serious disease such as cancer. This is often a vicious circle because as the symptoms of perimenopause increase then (without an understanding about what is happening to the patient), the patient's anxiety increases. This anxiety can even develop into depression in some patients, and once again, the signs and symptoms of depression are described earlier in this book. Some researchers suggest that up to 20% of women in the perimenopause may develop clinical depression. Very few, if any, patients will seek medical advice at this stage either for their perimenopausal symptoms or for their resultant depression. This will result in a considerable reduction in the quality of life of the patient, her family, her friends, and sometimes even her co-workers. It is critical that if you are of menopausal age and are suffering from the symptoms of the perimenopause resulting in anxiety and even depression, then you should seek medical advice immediately. This problem will not 'go away', and you will not 'pull yourself together'. The first people (apart from the patient of course) to suffer will be the partner and the children of the patient. They will almost certainly not understand what is happening to you (unless they have been fortunate enough to read a book like this), and they may react very negatively to your symptoms and feelings. This, of course, just makes things worse for the menopausal woman and for everyone around her. If medical advice is taken early, then these problems can be controlled, and there will be much less suffering for everyone involved. It is interesting and encouraging to note that the risk of depression appears to decline 2-4 years following the final menstrual period, especially in those women who suffered depression during the perimenopause. This does not mean that the menopausal woman should just 'suffer' 2-4 years of depression in the hope that it will go away; please seek medical help. On a positive note, this reduction in depression suggests that it is not age-related or part of the aging process; it is simply a symptom of the perimenopause.

Social Factors in Perimenopausal and Menopausal Depression

Each woman has her own social context, which may impact the likelihood of suffering from depression during the perimenopause. It has been suggested that single women, whether they be divorced, widowed or just single, are more prone to depression during the perimenopause. This may be because these women do not have the support of a partner during this difficult time and have no one they can trust with which to discuss their experiences and emotions. This does not mean that all single women will be depressed during the perimenopause but that they seem to fall at a higher risk of suffering from depression. This same group of single women, if they are also suffering from financial hardship, also seem more prone to depression during the perimenopause. Isolation and financial hardship are, of course, classic causes of depression in the general population, but most researchers agree that when these factors are combined with the perimenopause then depression is the likely outcome. There is no information at the time of writing to suggest that race or ethnic differences have any impact on the likelihood of developing depression during the perimenopause.

Other social factors which may increase the likelihood of depression during the perimenopause include such things:

- Sudden or long-term stress, *e.g.*, a long-term illness or a death in the family.
- Daily annoyance which can build up over time, *e.g.*, childcare or dependent relatives.
- Poor social relationships, *e.g.*, those women who may live an isolated life through choice or circumstances.
- Major events happening to family members, *e.g.*, weddings or children leaving home

In summary, in the social context, it seems that any adverse life event can impact badly on the perimenopausal woman and the likelihood of her developing depression. In terms of what can be done to reduce the risks of this depression, these include:

- Having someone to help with housework and other daily routines
- Closer involvement of friends, relatives and organisations (*e.g.*, women's groups)
- Someone to confide in and talk to, *e.g.*, friends, a healthcare worker, support organisations such as the Samaritans in the UK

All of these suggestions basically relate to good emotional support for the perimenopausal woman. The number and quality of this emotional support are critical to minimise the chance of depression, or even severe depression, developing.

Psychological Traits and Perimenopausal and Menopausal Depression

We all have psychological traits; mine are many and varied! Psychological traits which may impact the perimenopausal woman include:

- Pessimism (thinking that bad things/events are going to happen without any real basis).
- Anxiety or neuroticism (emotional instability) existed before the perimenopause.

- Rumination, *i.e.*, thinking about things or events over and over without a break, is perhaps associated with self-blame.

It is also interesting that if a patient has a negative attitude towards the menopause (which may be related to existing pessimism and neuroticism), they are also more prone to depression.

Psychological Adversity as a Child

The bad experiences some children suffer as a child have long been recognised as a trigger for psychological problems later in life. This is not to say that everyone will have suffered childhood adversity and that it can be 'blamed' for thoughts or actions in later life, but it is still nevertheless a fact that the two seem to be linked. Examples of childhood adversity include general family problems, poverty, abuse (physical and sexual) and unsafe environments. Most countries work as hard as possible to reduce or eliminate child adversity, but it sadly still very much exists. If child adversity is experienced by a woman who undergoes perimenopause and menopause, then she *may* suffer more severe psychological symptoms. This is something that should be discussed with a counsellor if it is important in your own situation.

'Brain Fog' in the Perimenopause and Menopause

When 'brain fog' was first described by perimenopausal and menopausal women, it was often dismissed by the medical profession because there was not enough clear evidence to prove that it exists. More recently, research has shown that brain fog in the perimenopause and in the menopause is a real symptom that should be taken seriously, and support is provided where needed. Many women describe 'brain fog' as a feeling of difficulty in concentrating, general forgetfulness, and an inability to think with any amount of clarity. The forgetfulness can be particularly worrying as some women may report a sudden inability to remember the names of people they have worked with for many years! These are, of course, especially frightening symptoms because they are also found in the early stages of many of the neurodegenerative diseases, such as Alzheimer's disease. In most women, it seems that brain fog decreases with time and does not seem to be related to the onset of dementia in later life. Nevertheless, brain fog is a very distressing symptom of the perimenopause and menopause, and if it becomes a problem, then do not seek to ask for help.

The Cause of Brain Fog

There is a considerable amount of ongoing research into perimenopausal and menopausal symptoms such as brain fog. The current thinking is that brain fog is directly related to the level of oestrogen, which of course, declines during the perimenopause and menopause. Oestrogen is responsible for the correct operating of a region of the brain called the hippocampus. The hippocampus is an area of the brain which is responsible for memory and thinking. It is thought that a decline in this activity in the hippocampus because of declining levels of oestrogen is the underlying cause of brain fog.

There is currently no known treatment for brain fog, but there are a few things that *may* be beneficial, such as:

- A healthy diet that excludes processed, fatty, salty and high-sugar food. There are so called 'oestrogen rich' foods such as sesame seeds, flaxseeds, soybean products, alfafa, olives and olive oil, chickpeas and multigrain bread. Even red wine has been suggested as being helpful, in moderation, but red wine has been touted as a panacea for many diseases in the past with little conclusive evidence.
- Maintain good hydration. Almost everyone seems to walk around with a water bottle these days so this should not be too difficult to achieve. Dehydration has been proposed to 'shrink' the brain, but there is no real evidence to support this idea, especially in terms of brain fog. Despite this, proper hydration does lead to better general health so it is good idea that might help to reduce brain fog.
- Reduce 'belly fat' if obese. High levels of 'belly fat' ***may*** be associated with dementia and Alzheimer's disease in later life. In general terms, it is good to reduce or remove belly fat as it may also contribute to other more serious diseases such as cancer and type II diabetes. Some researchers suggest that reducing weight can improve the memory although much more research is needed to confirm this claim.
- Ensure that you have a good sleep pattern with good quality of sleep. During sleep, the brain 're-sets' itself, and this process may be very beneficial to brain fog sufferers.
- Another thing which seems very popular is exercise. This does not have to be marathon running, but 30 minutes of exercise (*e.g.*, a walk) will benefit everyone, including those suffering from brain fog. Many people talk about exercise as 'clearing their head', and it is certainly something that can be done by anyone and may be very beneficial in many ways.
- Some brain fog sufferers report that meditation can helpful. This will, of course, not apply to everyone, and not everyone has the mindset for meditation. Nevertheless, it might be worth a try since the level of relaxation reached during meditation may reduce stress which in turn may reduce brain fog.
- Acupuncture may be useful, but this is probably more related to the associated stress reduction rather than a specific 'treatment' of brain fog.

All of these suggested lifestyle and activity changes may help some people suffering from brain fog and there may also be other benefits to the general health.

CONCLUSION

The psychological impact of premature menopause, perimenopause and menopause can be serious and long-lasting. Not all menopausal women will suffer such symptoms, but the more the menopausal woman understands her symptoms, and what can be done to reduce or alleviate these symptoms, the better. It must be remembered that the menopause is a natural physiological event which happens to all women across the world whether they are rich, poor, famous or even infamous! In some cultures, the menopause may even be considered a 'taboo' subject that is neither acknowledged nor discussed. This can result in much suffering for menopausal women. Other cultures view the menopause as a positive event, such as in China, where the menopause is often referred to as 're-birth' or 'second spring'. In Japan, the menopause is considered to be a natural life-stage (which is exactly what it is) associated with energy and renewal. In Thailand, there is even a formal ceremony for women who have gone through the menopause with respect and rejoicing. I have worked for many years with colleagues in India, where menopause is thought of as a natural process which means that the menopausal woman takes a much greater role in such things as family decision-making, providing advice to younger people, and taking part in community events.

There is a difference between East and West in the way people think about and regard menopause, and it seems that the West still has a lot to learn!

The psychological and physical changes during the menopause are unique to each woman, ranging from no symptoms at all to severe life-changing symptoms. Physicians are aware of this, and patients must understand that they have to seek help if they are affected by serious symptoms. The correct level of understanding by patients and the appropriate help and support from healthcare professionals will enable all women to undergo the menopause with a positive outlook on life. This is no more a taboo.

KEY POINTS OF CHAPTER 7

- The menopause presents not only physical but also psychological challenges to all women.
- Physical and psychological problems related to the menopause do not apply to every woman; each woman is different in her experience of the menopause.
- Premature menopause is a life-changing diagnosis for a young woman, which *may* result in ongoing stress or psychological problems.
- Society is becoming increasingly aware of mental health in general and

specifically how mental health can change during the menopause. Any woman (or man) experiencing mental health issues of any kind should seek medical advice immediately.

- Fear is a common emotion in menopausal women. If this happens, please seek medical advice immediately.
- Single women may be more prone to stress and anxiety during the menopause, especially if there is no immediate family to provide support. If you find that this is a problem for you, then please seek medical advice immediately.
- 'Brain fog' is a very distressing symptom of menopause which, if it occurs, can be extremely stressful. If you start to find difficulties in clear thought and memory, then please do not sit and worry. Please consult a medical professional as soon as possible.

CHAPTER 8

The Media, Celebrities and Menopause

(Help or Hinderence?)

When it's down and dark, small things become ridiculous; when you don't find joy in life, it's probably when you are in the menopause, or you may be seriously depressed for other reasons; it's tiring and worrisome to be seen as old, odd or cranky and not fitting into society. Facing the fact that you no longer can have children is sad. The way forward is to realise it is a phase, unavoidable, and to share it is incredibly empowering

Helen Lederer

INTRODUCTION

The media and the internet are constantly present in our daily lives, driving astonishing levels of marketing to the global population on a 24/7 basis. When I was born in 1958, none of this existed; Some may say this was a good thing. Others may say that information and knowledge are empowerment and critical component of modern life. I perhaps sit somewhere between the two opinions. Every person on the globe seems to have a phone in their hands and 'checks' it sometimes hundreds of times per day. In our 'connected' world, wars and disasters are reported in graphic detail, and the latest 'fad' or 'must have' is presented to everyone on a regular basis through multiple formats. This results in people 'wanting' things that they may not actually 'need', especially in developed countries. The marketing industry which drives this information barrage is currently estimated to be worth $1.7 trillion and growing. The 'perfect life' is portrayed in a way that does not actually reflect the life or experiences of most of the population on planet Earth. Health issues are not exempt from this information and marketing barrage. Discussions about specific diseases, treatments (often at high cost), and advice are common, and are often, at best unwelcome and, at worst, potentially dangerous. The positive side of all this interaction is that it will at least get people talking and thinking globally about disease and treatment, which would have been very unlikely before the 'information revolution'. The solution to this problem (as in many things in life) is quality, not quantity. One page of accurate, clear information is much more valuable (in information, not in money) than a thousand pages of misleading, incorrect and potentially dangerous information. Everyone has a Duty of Care and a Duty of Candour from the healthcare professionals who carry out their diagnosis of treatment. Duty of Care means that nothing must be done which will harm the patient. The duty of candour means that all healthcare professionals must speak the truth when talking to patients. Some media people will not even know that such standards exist and would do well to consider opting for such open, clear and honest behaviour.

THE MEDIA AND THE MENOPAUSE

The general attitude to the menopause in society is in a very slow process of change. In some populations, it is, for some strange reason, a taboo subject. This may be due to fear from patients and an overall poor understanding of what the menopause really means. Despite all of this, menopausal patients are slowly being managed in a better way by healthcare professionals, and the information available about menopause is increasing, and in general, terms is reliable. There is no doubt that there has been a steady increase in the frequency of articles about menopause (online and in hard copy) not only in medical journals but in general books (such as this), websites and magazines. This is a good thing. Nevertheless, the media still has some way to become more effective and clear in promoting the understanding and empathetic mindset needed in menopause. Improvements could be made in the following:

- The increased discussion of the menopause in the media is to be encouraged, but sadly some of this information is minimal and sometimes insufficient, leaving the menopausal women themselves thirsting for more information. One way to improve this situation would be for better collaboration between the media, the experts in menopause, the relevant menopause charities, and the menopausal women themselves. This would result in more accurate information and much greater empathy in the discussions.
- The media often portray the menopause as a disease and a negative experience in a woman's life, or sometimes a taboo subject not for discussion at all. Some even suggest that the menopause is a disease that needs medical treatment in the same way as heart disease, for example. Menopause is not a disease; it is a natural, normal physiological process. Equally, it is not (hopefully) a negative experience, but sadly for many women, the menopause can be a very negative experience. The symptoms of menopause suffered by some women are certainly unpleasant and unwelcome and need treatment where possible. The negativity promoted by some areas of the media simply makes menopausal women feel worse, and this is unacceptable in a modern democratic society. Imagine the harm done if the same attitude was taken to stroke, heart disease or cancer! Menopausal women need information from the media, which is helpful and supportive, and there is no reason why this cannot be achieved.

- The media may sometimes get the description of the menopause wrong and the treatment advice incorrect. This is simply down to poor research, and collaborations, as suggested above, would solve this problem immediately.
- The discussion of the factors associated with a 'good' and a 'bad' menopause is often misrepresented. Important factors, such as exercise and diet, race and ethnicity, stress and lifestyle, are often ignored or trivialized. These are critical components of the menopause and, where possible, may minimise symptoms if managed correctly.

Menopause is rarely mentioned in the media, and when it is then, it is usually in an over-censored or controlled way. A feeling of fear (probably based on poor understanding) underlies such information, which needs to change to help all menopausal women. A menopausal woman is most definitely *not* 'past it' or 'over the top' or has 'suffered the change', and the subject of the menopause must not be discussed in whispered or frightened language. The subject of menstruation is often discussed in the media in a factual and honest way; why cannot this be the same for the menopause, which is the natural progression from menstruation? There may be an opportunity here for better education at secondary and higher levels. When I was in school (admittedly a long time ago), there was a slight nod to menstruation in biology lessons and nothing on menopause. When I studied at Cambridge University, there was considerable information provided on all aspects of female reproductive physiology but not a great deal on the menopause. When I was lucky enough to join the team who developed *in vitro* fertilisation (test-tube babies) at Bourn Hall Clinic, menstruation was our daily work! Equally, during these times, there was little mention of the menopause apart from saying that if the patient was perimenopausal, then IVF would be extremely unlikely to work. This is an opinion which sadly still stands today. Our research described in Chapter 6 might change this in the future.

The media could, in fact, play an important role in highlighting treatments for menopause which could make some patients realise that the treatment they are on may not be right for them. The media clearly has a powerful influence on the behaviour, attitude and beliefs of social groups and individuals in any context. Such an influence might be very powerful in educating and informing about the menopause if it is used correctly. Media bosses, please take note!

Celebrities and The Menopause

The power and impact of female celebrities openly and honestly talking about their menopause experience is a powerful and extremely important part of menopause education. Such celebrities are generally 'well known,' and people relate to them on many levels. Female celebrities, from actresses to television presenters to a previous First Lady of the USA, have all very openly talked about their own experiences of the menopause. This openness and honesty are extremely valuable. Such celebrities may also associate themselves with a particular menopause charity, which is a very positive way to help all women going through the menopause.

Most recently, there have been two notable television documentaries by female celebrities in the UK, which have had a very positive effect on the understanding of the menopause. These were:

• The Truth About The Menopause (featuring presenter Mariella Frostrup)
• Sex, Myths and the Menopause (featuring presenter Davina McCall)

Both of these were great, inspiring pieces of television and promoted considerable interest and discussion at the time of broadcast, but it was a short-term effect. Both presenters did 'follow-ups' on various chat and news shows, but sadly, these were once again transient events. Other things happen, and people forget quickly and move on to the next 'exciting' or 'interesting' thing. If you are able to view either or both of these documentaries, you will see that they both have a similar format. They were presented by prominent celebrity women, and they talked about their own experiences and those of other women from the general public. The clinical data discussed was excellent and correct in both programmes, with expert opinion sought where appropriate.

Not surprisingly, the message from both documentaries was that menopause is an area full of myths, taboo, confusion and misinformation. Even the Women's Health Initiative seemed to have had little impact in improving care for menopausal women. Menopause specialists on a global basis strive towards best practices in the treatment of the menopause (and this was emphasised in the documentaries), but the overall message is still failing to get through so that menopausal women all get the treatment they deserve.

There are some critical needs still outstanding to ensure that menopausal women get the correct and appropriate treatment. This is bearing in mind that the menopause experience is almost unique to each woman, so a 'one cure fits all' approach is particularly useless in the treatment of the menopause. The intervention of 'personal medicine' may in the future make the 'one cure fits all' concept obsolete; sadly today, the concept still holds in routine clinical practice. These needs are:

• Good, reliable, up-to-date, clear, and honest information available, when needed.
• Availability of this information in as many formats as possible (many people,

for example, do not have access to the internet on a global basis or have the time to read books like this).

- Information which can be applied to individual cases, *e.g.*, if everyone provides information on hot flushes, but no one provides information on brain fog, then such a lack of thorough coverage of symptoms may just lead to further frustration for the patient.

It is true that many physicians struggle to provide the most appropriate treatment of the menopause because of the complexity of the condition. Nevertheless, the same physicians might cope very well with the treatment of heart disease, cancer and stroke. It is also essential for the healthcare providers treating menopause to be up to date in their knowledge. Many might not have the time or inclination to get involved in Continued Medical Education (CME) related to the menopause. In addition, we have enormous amounts of excellent menopause research, which may just stay in the journal if investment in clinical trials is lacking.

KEY POINTS OF CHAPTER 8

- The internet and social media must be judged with a critical mind. The information on the internet is often simply untrue, especially in 'chat rooms' or social media. This can be harmful to everyone and can be very damaging. Please use the internet selectively and with care.
- The media have an important role to play in healthcare education, but only if this role is supported and approved by the appropriate experts. Some media reports should be viewed with a level of skepticism, and those in menopause are no exception.
- Celebrities and other 'famous' people may have an interesting and helpful insight into the menopause. Once again, such information and discussions must be viewed critically, as they could be biased or incorrect. Nevertheless, there are some inspirational people who talk openly, fairly, and honestly about the menopause, which can be very helpful in the overall discussion.

CHAPTER 9

Family, Friends and Work

(Menopause Support)

The family is one of nature's masterpieces.

George Santayana

INTRODUCTION

When someone suffers from any disease or condition, then that person always benefits greatly from the support they get from their family, friends and work colleagues. This is no different in the case of menopause, and the support received from these key people will make a massive positive difference to the women undergoing the menopause. These human-to-human interactions are often taken for granted, but in times of stress (be it medical or psychological), they are critical to a good outcome for the patient.

FAMILY AND FRIENDS

Families and friends can be extremely diverse in geographical context; some live on the same street, and others, like my family, are spread across the UK and Canada. My friends and colleagues can be found across the globe, from the USA to India. This diversity of how friends and families are placed geographically makes it difficult to generalise the role of family and friends in supporting women undergoing the menopause. Simple time and distance may make simple and effective communication difficult. There are, of course, numerous meeting platforms on the internet, which I use almost every day, but I must admit this is mostly for business. These are a great way to stay in touch with family, especially on a global scale, but they can never be a substitute for a hug from someone you love. We all discovered this during the COVID-19 pandemic.

In terms of the menopause, the support of family and friends, to help and reassure you, may be just as important as the medication or other treatment you receive. The peace of mind from knowing that someone loves you, and will support you no matter what, is an extremely powerful thing to have in any form of adversity or stress. The menopause is no exception. These interactions with friends and family are not only about support, but they may also result in new ideas or methods in which you can reduce the symptoms of the menopause. To be effective in their support, family and friends need to make an effort to truly understand and appreciate what you are going through during the menopause. This might be tough for some people, but a little effort could make a big difference to your family member or friend going through the menopause. There are some key things that family and friends can do to help anyone going through the menopause, these include

- Finding out about the menopause and the symptoms it can produce. The more you understand what is happening to your family member or friend then, the more helpful you can be. This knowledge can be found in books such as this or by using our old friend 'the internet'. If you do choose to use the internet to find out about the menopause, then please make sure that you look at reliable sources such as those from menopause support groups or those written by physicians.
- Perhaps the most useful thing a friend or family member can do to help a menopausal woman is to talk to her (using the knowledge you have gained above). This will show her that you understand, that you love her and that you will do your very best to take care of her. This simple conversation could be life-changing for a menopausal woman. All patients need love and support during any illness or accident, and the menopause is no different. Menopause is not taboo, it is as important as the help you would give to someone with heart disease or cancer.
- The most difficult aspect of the menopause for friends and family is the raw emotions that can be shown by a menopausal woman. She may be depressed, she may be sad, she may even be angry, she could even be suicidal. Handling these emotions requires very high levels of patience from friends and family. In the case of severe depression (as described earlier in this book), which may or may not include suicidal thoughts, healthcare professionals must be consulted immediately. This is a life-threatening situation in the same way as severe trauma, a heart attack or a stroke. Please contact the emergency services and get help as soon as possible. This rapid and effective action by those supporting the patient is needed to ensure that she remains safe. The words which are used by friends and family are equally important: 'Pull yourself together' will not work, neither will 'see how you are in the morning'. Some menopausal patients in this extreme situation might not make it until the morning. For the less stressed menopausal woman, who is nevertheless suffering, words such as 'this is not going to last forever' may provide some comfort as she may realise that this is just a stage in her life and not a permanent change. Throughout all of these interactions, family and friends will need to have very high levels of patience.

The menopausal woman may seem irrational or 'not like she used to be,' but with patience and love, you will all get through this difficult life change unscathed.

- Most menopausal women will have moments of extreme upset. This may materialise as sadness, anger or frustration. This upset may be triggered by what may appear to be trivial matters in the eyes of friends and family, but they are very stressful to the menopausal woman. If it is possible, friends and family need to try not to 'mirror' these emotions as this will lead to conflict between you and further unnecessary suffering. The best advice to friends and family here is to listen to the what the menopausal woman is saying without getting angry, sad or frustrated. Being calm and listening without criticism or unwanted emotional involvement will be enormously beneficial to the menopausal woman who finds herself in this situation. This is a difficult thing to ask of friends and family, but if it can be implemented, then great benefit and support will be achieved.

- The menopausal woman can often feel fragile to the point where she may not even 'recognise' herself because of the turmoil the menopause brings. She may say that she is 'doing her best' and it is important for friends and family to support this notion. This is because the menopausal woman is actually doing her best. This might not be what she did before the menopause, but if friends and family believe in her and say 'yes, you are doing your best and we all appreciate your efforts' then this will be very re-assuring to the menopausal women. Show love, show understanding, show compassion and the transition through menopause and beyond will be very much smoother.

- The menopausal woman will often feel overwhelmed by the daily tasks, jobs and chores she has to do on a daily basis. This may be housework; it might be driving, it might be an employment-related task or even something such as making dinner or organising a birthday party. The key thing friends and family can do here is offer help. The menopausal woman may reject this offer of help, but if she does, then re-emphasise the offer because your help may be extremely useful in relieving the feelings of pressure and lethargy suffered by many menopausal women.

- Good quality sleep is essential for everyone, but in the menopause, insomnia is a very common symptom. If the insomnia is severe (*i.e.*, getting less than 8 hours sleep per night), then medical help is needed immediately. Short-term sleeping tablets or relaxants may be prescribed. Most menopausal women find that the insomnia can be reduced by making sure that the bedroom is well-ventilated and at the correct temperature (16-19°C), and that the lighting is optimal (a black-out blind is often useful when the dawn is in the early hours of the morning) and perhaps most importantly the bed and pillows are comfortable. Most people have beds that are very old and uncomfortable. This does not help anyone suffering from insomnia! There are also relaxation techniques which some women may find extremely beneficial to enable good sleep. An electric fan might be helpful for those hot-flushes and night-sweats. Good sleep will mean good (or at least improved) mood, so this is in the interests of everyone. It is most important for menopausal women to get a good sleep every night, if this becomes impossible for whatever reason, then please seek medical help.

- The better the general physical and mental health of a menopausal woman, the better will be her transition through the menopause. Friends and family can help by suggesting an exercise plan which will be completed together (nothing too strenuous!). The amount of benefit a menopausal woman gets from such activity and support is very significant. It may be a gym-based work-out programme, it may be any outdoor activity such as golf, sailing, hill-walking or even a gentle walk on a Sunday afternoon. Such activity is proven to improve mental health, and it may even help to lose a little weight which is always beneficial. The key thing here is to 'cajole' the menopausal women into joint activities rather than saying, 'right, we are now going for a ten-mile walk'. Gentle encouragement, support, and even praise will not go amiss.

- Many menopausal women say that their self-confidence has been taken away. Visual changes to her skin, weight gain, hair changes (volume and colour) and even incontinence can significantly impact her self-confidence. Women once known as being flamboyant and interesting, can become withdrawn and introspective. In terms of self-confidence, friends and family can re-assure the menopausal woman still has enormous worth and importance within the family and that their love for her does not change just because of menopausal symptoms.

- Most menopausal women feel some level of anxiety and guilt. The anxiety and guilt may relate to the feeling that she has 'changed' and is no longer as important in the family unit as she was before the menopause. Looking at it from the outside, this is patently untrue, but in the mind of a menopausal woman, the feelings may be extremely powerful. Doubts and irritability in the mind of the menopausal woman will only escalate the feelings of anxiety. Support and reliability from friends and family will help a menopausal woman to handle these complex emotions.

In summary, the friends and family of a menopausal woman carry out a critical role in the way in which a woman experiences the menopause. It is, however, important to remember that the menopause is a different experience for almost every woman. Some may need considerable help and support from the family, others may need very little. This requires some level of judgement by friends and family, and the guidance given above is a good starting point. It is important not to be over-demanding on the menopausal woman, but at the same time, she may often say that she is reluctant to do things that eventually prove to be extremely beneficial. Most important of all, give the menopausal woman love, friendship, trust and laughter. Make her realize that this is the second stage of her life which can be just as much fun as the first!

The Role of the Husband or Male Partner of the Menopausal Woman

Undoubtedly, the husband or male partner of a menopausal woman will be the most affected family member. He needs to have an extremely high level of understanding, compassion, and love for his wife or partner. He may feel betrayed, he may feel unloved, he may feel frightened, and he may feel that his marriage or partnership is falling apart. Worst still, he may be suffering from the andropause at the same time as his wife or partner is suffering from the menopause! This sounds like a recipe for disaster, but with the correct level of understanding and support, the male partner or husband of a menopausal woman can be the difference between imminent disaster or future harmony and love in the next phase of their lives. There are several areas which the husband or partner of a menopausal woman needs to be aware of and have an understanding, these include:

- You, your wife or your partner may lose interest in sex. This could be a gradual process, or it could just happen one day. The first thing to say is that this has nothing to do with the male partner or husband and that he is in no way to blame. This behaviour of your wife or partner is due to the decreasing levels of the hormone oestrogen which drives the female libido. The menopause can make penetrative sex for your partner or wife painful due to vaginal dryness. It is critical that you and your wife or partner discuss this and perhaps explore other types of intimacy which are not uncomfortable or painful for your wife or partner.
- It is important to try to maintain a sense of humour when appropriate, but this can be extremely hazardous in the wrong situation or context. Proceed with great caution and be ready to sincerely apologise if you get it wrong. Remember when you first met and how you made each other laugh. If this kind of interaction can be re-kindled, then this may make your wife or partner feel more relaxed, reduce her anxiety, and increase her self-confidence.
- There will inevitably be times of upset; even the very 'best' menopause will have ups and downs. In times of upset, try to focus on the good times, such as holidays or when children were born (assuming that this was not too traumatic!) or other achievements made by your wife or partner, such as in sport, fitness, music or anything which was positive and rewarding. There may also be times when you worked well together as a couple, and reflecting on such events could be very beneficial when your wife or partner is going through an upsetting time.

Use your imagination. You know your wife or partner better than probably anyone else in the world. Use this unique knowledge about your relationship to restore happiness and harmony.

- Your wife or partner is certainly going through a difficult time when she is undergoing the menopause. Even an 'easy' menopause can raise serious problems which have to be tackled. It may, of course, be possible that you are also going through a difficult time, such as andropause, work issues, financial problems and caring for dependent children or parents, and so on. We all have these complications in our lives, and the way in which we handle them is critical to a happy and prosperous future. Nevertheless, when these things all happen at the same time, then the overall problems may seem to be magnified several times and may even seem intolerable. This kind of scenario may require additional external support, such as the GP or Family Physician, counsellors and advisors, to help you both through these extremely difficult times. Do not hesitate to seek help if you need it or if someone else suggests that you should seek help (they may see what you do not). It is extremely important that both you and your wife or partner talk about these problems. This is easy to say but often quite difficult to do, especially if you have been drawn apart by the issues you face. I find talking whilst walking (especially hand-in-hand as this gives an immediate physical and emotional connection) to be somehow easier to do, give it a try! If you do try this method, be sure to choose a nice day; walking in the cold and rain is generally not good for the mood. Remind yourselves that balance will return, and the current distorted and uncomfortable situation *will not last forever*.

- A menopausal woman clearly needs a lot of support from her friends, family and most importantly, her husband or partner. This support must be in the form that your wife or partner will appreciate and make them feel a little better. The key here is that the support you provide must be what your wife or partner *asks* for, not what *you* decide she needs. For example, your wife or partner may want a quiet, intimate evening together, but you decide that she needs a large party with as many people as possible. Ask, listen, and respond accordingly. It might not be what you want but it will certainly be what she wants and needs.

- Whilst your wife or partner goes through the menopause, there will be times when you feel angry, frustrated, at fault, rejected, and generally very unhappy about the whole process. This is inevitable and something the male partner should be aware of and brace himself for. These feelings are not your fault, but they can be extremely damaging if allowed to persist in your mind. It is a big ask, but try to control these feelings and emotions. They are an unwelcome by-product of what your wife or partner is going through.

- Following on from, and re-emphasising the point above, it is worth reminding the husband or partner of a menopausal woman that the situation and emotions expressed must not be taken personally. This could result in an unnecessary conflict which is not beneficial for either person.

- When in discussion with your wife or partner, please try not to dismiss the situation you find yourself in as 'hormonal'. This may be interpreted as a dismissal of the suffering your wife and partner are going through as trivial. She needs your respect, understanding and love more than at any other time in your relationship so far.

- Following on from discussions, it is very important to choose your words carefully and to avoid any 'opinions' which may not be evidence-based. A common thought amongst male partners or husbands of menopausal women is that: 'There must be a cure for it'. Modern medicine is extraordinary, and new ideas and treatments are evolving all the time for many diseases. In truth, there is no current 'cure' for the menopause, but certain things may help with the symptoms. The main treatment of the symptoms of the menopause at present is Hormone Replacement Therapy (HRT). This may help many women, but it does not work for everyone. Other treatments may be tried as described earlier in this book, but once again, these will not be a 'cure'. The basic message here is a suggestion that there is a 'cure' for menopause is incorrect and best avoided in discussions.

- It is also important to avoid suggesting that your wife or partner should seek to be prescribed anti-depressants or anti-anxiety medication. Unless you are a physician with considerable experience in the treatment of menopause and related symptoms, then such a statement is totally inappropriate. Your wife or partner can seek such medication if she feels it is appropriate; this must be her decision, not yours. The process should not be under pressure from you. The only exception to this is if you feel that your wife or partner has become seriously depressed. The symptoms of serious depression are described earlier in this book. If you notice any of these symptoms, then it is appropriate for you and other family members to ask her to seek medical help. If she refuses to seek help, then you and the family may choose to alert her GP or Family Physician, who will make the appropriate contact with your wife or partner.

- There is no 'quick-fix' for the menopause. In your role as the husband or partner of a woman going through the menopause, you need to find out what she needs, listen to her (*properly* listen to her), not dismiss her opinions and needs, and most of all, be supportive. A menopausal woman who feels that she has no support from her family and friends will be in a very lonely and vulnerable place.

- Your involvement in terms of practical help will be very welcome. This may be house cleaning, food preparation or anything where your wife or partner feels too tired to cope with this work. You may, after discussion and approval from your wife or partner, take over the responsibility for some tasks. This may take a lot of pressure off your wife and partner, but it must be done with her approval.

In summary, your wife or partner going through the menopause needs love, more love and then a little bit more love. She also needs support and re-assurance despite her apparent unusual or erratic behaviour. She does not want criticism or, worse still, angry silence. If you yourself feel stressed about the situation you find yourself in, then please get in touch with your GP or Family Physician and explain what is happening. The point of all this is that you both get through the menopause and go on to a loving, happy and healthy 'second 'stage' of your life together.

The Single Woman and The Menopause

Interestingly, single women often report an earlier onset of the menopause than married women. No one really understands why this might be. Being single and going through the menopause can, for some single women be an opportunity to embrace the process and treat the symptoms resulting in a perceived feeling of 'liberation' and 'control' of her life. This may be because a single woman has the maturity and freedom given by being single but also because she can deal with the menopause in her own way without the pressures which may arise from a partner or husband. A single woman, of course, has her extended family who may provide support, especially sisters and other younger and older female relatives.

Being single can allow a woman to ignore the stereotypes and media images of the menopause and come to her own decisions, actions and treatment. As for the married woman, or the woman in a relationship, it is essential to view the menopause as a natural progression from puberty and not a 'disease'. Many women feel that the menopause can give a significant transformation to their lives and that they can look and feel better after the menopause. The process itself may be difficult, but the time following the menopause can be when a woman looks and feels better than ever. This process takes a little help (for example, from healthcare professionals) and also some self-care which may be applied with much more freedom when single. The mid-life years for the single woman may be full of vibrancy and new experiences. The most obvious benefit of the menopause is that monthly periods no longer exist. These can often be very painful and debilitating for some women, and a reprieve from this monthly event is often greeted with great happiness. This can very easily result in a better social life and, of course, no risk of pregnancy. A note of caution here: pregnancy is still possible in the perimenopause!

NEW ACTIVITIES

It may sound a little cliché, but the menopause is truly the start of the 'second half' of a woman's life. The menopause may start as early as age 40, and there are at least another 40 years to live for the average woman. This 'second half of life' should be viewed just as happily and enthusiastically as the first half. The opportunities, interests, excitement and love of life do not stop because menstruation stops. This may be the time to learn new activities and skills; it might be even the time to start a career or job which might have been impossible or impractical in the first half of life. This new level of maturity may also help to eliminate any insecurities or worries which may have been present during adolescence and adulthood. In short, the post-menopausal time could be a time of great personal fulfillment and renewed self-confidence to try and experience new things. The menopause is not all doom and disaster; there are many positive aspects which should never be ignored.

WORK AND THE MENOPAUSE

Most women undergoing the menopause will be in some sort of employment which in some cases may be critical to their financial well-being both for them and their families. Other women may be in high-pressure jobs such as medicine, where a lack of clarity of thought or sheer tiredness could be dangerous to others. Therefore, all employers must be aware of menopause and understand that it may affect the ability of their staff to carry out their work confidently, safely and effectively. It could, of course, be that an employee is not the person actually suffering from the menopause. It may be a person helping a woman through menopause, such as a partner, relative, carer or even another colleague in the same workplace. This person could be highly stressed both emotionally and physically and his/her own work may also decline. It is essential that people in the workplace, including men, should be involved in training and conversations about the menopause so that they can help in the workplace. The risks, behaviours and symptoms that menopausal women face in the workplace will vary, but in general terms, employers should be aware of:

- An employee suddenly lose their confidence and their previous skills and abilities decline.
- An employee who is suddenly taking more time off work (either holiday leave or sick leave) but hiding the reason behind their absence.
- An employee with decreasing levels of mental health, which may appear as increased stress levels, anxiety, or even depression. This is impossible for an employer to properly assess but may be seen in changes in general behaviour.
- An employee who leaves her job for no rational reason.

The employer cannot be responsible for the fact that an employee has started to go through menopause. Nevertheless, the employer (probably led by the Human

Resources Department if one exists) could ensure that the following are in place:

- It is important that the employer has policies and procedures in place to manage and support menopausal employees. This will be relatively easy in large employers, but smaller employers may struggle. In this event, a small employer should take advice from many of the menopause charities on how best to support their staff.
- Any existing policies and frameworks relating to the menopause must be regularly reviewed and revised to ensure that the support being offered is current and valid.
- It is important that employers break the 'silence' or the 'taboo' surrounding the menopause. This is done by training and discussion and would ideally be part of mandatory training.
- Those employees suffering from the menopause and experiencing significant symptoms should have reasonable adjustments made in their work pattern to reduce stress.
- Line-managers need the ability and resources to support their teams. This is not only for the menopause but for all other medical conditions which may effect the way in which an employee can work.
- Senior management within the company must be part of the overall management of menopausal staff to ensure that the correct level of support is being provided.

There are several online resources for both employees and employers in relation to the menopause, which is listed at the end of this book.

KEY POINTS OF CHAPTER 9

- Family and friends are an important source of love, support, and in some circumstances, advice (*e.g.*, a female friend or family member who has already been through the menopause).
- The husband or male partner plays a critical role to supply in being the most loving and supportive person a menopausal woman can turn to. This is a big demand on male partners which some may find difficult to achieve. Reading this book and similar sources of information may be helpful.
- The single woman may feel anxious and isolated during the menopause. Friends and family are once again an important resource for single women.
- New activities and interests may develop post-menopause in the 'second-half of life'.
- Employers and work colleagues should, if possible, be aware of the problems a menopausal woman may be suffering. Many employers now claim to be 'menopausal friendly'.

CHAPTER 10

Lesbians

(Menopause and the Lesbian Woman)

My feelings for Ellen overrode all of my fear about being out as a lesbian. I had to be with her, and I just figured I'd deal with the other stuff later.

Portia de Rossi

INTRODUCTION

The biological process of the menopause in a lesbian woman is exactly-the-same as the menopause in a bi-sexual woman, or in the menopause in a heterosexual (straight) woman. This statement, once again, makes this chapter look like it is going to be very brief. In actual-fact, there is quite a lot to say about the menopause in lesbian women, which can easily be seen from the general literature and the medical and scientific literature. This is a serious subject about which all relevant healthcare professionals should be aware. For example, when comparing lesbian and straight women in terms of the menopause, it is necessary to consider the age at menopause, the ethnic origin of the women and the education of the women. As in all scientific comparisons, it is important to compare 'apples-wit--apples' and to have one variable which is under investigation. In this case, we need to compare a group of women who have the same overall characteristics and then compare the variable between the two groups, which is their sexuality. This approach will allow a sound scientific basis for comparing whether-or-not sexual orientation has an impact on the menopause and any aspect of women's health. Sadly, such studies are currently few and far between.

THE MENOPAUSE IN LESBIAN WOMEN

As described earlier in this book, the menopause is very much 'individual' to each woman, whether-or-not she is lesbian or straight. Generalisations in this context (often based on straight relationships and not on lesbian relationships) are unhelpful and can indeed make some people feel worse or even develop fears which were, in fact, never there.

Lesbian menopausal women may also think that their feelings and views go unheard in the 'general debate' about the menopause. It is absolutely essential that the lesbian couple, when going through menopause especially, maintain clear understanding and supportive communication between themselves. This communication may be very different in heterosexual couples where the male partner has no 'understanding' of the feelings, emotions and physical symptoms of the menopause (unless he has read this book!). He cannot relate to the menopausal woman with perhaps the same love, understanding and compassion, which lesbian couple may be able to provide. Another big difference is that a menopausal lesbian couple may undergo the menopause at roughly the same time with roughly the same symptoms. This is the perfect situation, as mutual support will be much easier. Equally, each lesbian partner may quite likely experience the menopause at very different times with very different symptoms. If such an 'imbalance' occurs in the menopause in each lesbian partner, then it may put considerable strain on a lesbian relationship. One partner may feel relatively well whilst the other might feel relatively awful. This is a time when great understanding, love and tolerance is needed by both lesbian partners to ensure the sustainability and stability of their relationship.

Associated Risks of the Menopause in Lesbian Women

Earlier in this book, the risks associated with the menopause have been described, and at that point, I was considering only straight women. There are subtle differences in the risks when considering lesbian menopausal women. Some authors have suggested that lesbian women may be more prone to cardio-vascular disease and some types of cancer. The biological reason for this is unclear, but part of the reason may be due to life-style choices in lesbian women. This assumption about life-style has been confirmed by the Women's Health Initiative (WHI), which found that there is a higher incidence of risk factors such as alcohol consumption, smoking and obesity in lesbian women. These additional risk factors present in lesbian women may increase the incidence of menopause-related diseases such as cardio-vascular disease.

It has also been suggested that lesbian women are less likely to use healthcare services when compared to their straight counterparts. The reasons for this lack of trust in the healthcare system by lesbian women are worrying because this may mean that lesbian women might not be receiving the primary care they deserve. Some lesbian women may feel a possible lack of confidentiality or sensitivity by healthcare workers to lesbian women. If this is true, then there is clearly a very poor level of understanding of the menopause in lesbian women by some healthcare professionals. This can easily be resolved by re-training of healthcare professionals on this subject and more discussion of the subject in medical conferences and publications. If there is ever a lack of understanding and communication between the patient and healthcare provider, then this is a recipe for confusion (and even possibly disaster) in the treatment. This may apply especially to the treatment of lesbian menopausal women.

Common Misunderstandings in the Sexual Health of Lesbians

Some inaccurate reports show lesbian women have been exposed to the less sexual transmitted disease (STD) than their straight counterparts. This is incorrect and could contribute in part to the overall poor care of menopausal women. STD is a problem for all women and men. The incidence of cervical cancer, for example, is the same in lesbian women as it is in straight women. It is known that the majority of cervical cancer (in any woman) is caused by a virus called the Human Papilloma Virus (HPV). HPV is spread through sexual contact (homosexual or heterosexual) with a person who already has the virus. Being positive for HPV (in all women) does not always mean that cervical cancer will follow in fact, in some women, HPV infection results in genital warts. The risk factor for developing cervical cancer following HPV infection is increased by factors such as:

- Having sex at an early age.
- Multiple sexual partners.
- Smoking (in any context reducing or stopping smoking is a great benefit to the overall health of anyone).
- Infection with other STD's such as syphilis, gonorrhoea, HIV/AIDS, or the very common STD called chlamydia.

In addition, if a lesbian woman (or a straight woman) has never been pregnant then she may be at increased risk of developing breast cancer or ovarian cancer.

Domestic Violence and Anxiety

Domestic violence was on the increase at the time of writing, with approximately 23% of heterosexual females being assaulted by their partners at least once in their lifetime. This has been exacerbated by the COVID-19 'lockdowns'. These 'lockdowns' resulted in millions of vulnerable people being forced to stay with their partner all the time, increasing their risk. This risk applies to both heterosexual men and women. It is interesting to note, however that domestic violence is less likely to occur in homosexual relationships, especially in lesbian relationships. It is reported that physical aggression does occur in lesbian relationships, but physical violence is much less common when compared to heterosexual relationships. Despite this, lesbian women do experience homophobic abuse, which can cause considerable anxiety. Such homophobic abuse is totally unacceptable at any time, but most of all during the menopause.

Anxiety levels will increase and may even develop into depression, as described earlier.

Action Needed to Improve the Care of Lesbian Menopausal Women

Every woman going through the menopause (whether she is lesbian or straight) may suffer very different symptoms and also deserves the correct and understanding level of healthcare.

As mentioned above, healthcare providers and lesbian women may need more information and education on the possible lifestyle effects that can impact menopause. More work is needed, and studies carried out, on lesbian women in relation to the menopause. The bulk of the medical literature and information at the time of writing refers to straight women.

It is essential that a lesbian woman can talk about her sexuality freely, especially during the menopause, which is a very challenging time for any woman. Healthcare providers must be able to ask the right questions and, most importantly, fully understand and act on the replies.

A good start would be not to assume that the woman being seen by a healthcare worker is heterosexual. Such an assumption may cause a lot of confusion, and a simple question at the start of the discussion would clarify the situation. Healthcare professionals must not assume that being straight is the default.

CONCLUSION

What is clear is that the menopause can be just as complex in both the homosexual and heterosexual relationship, and the feelings and emotions of each group may be very different and difficult to handle. This might be thought of as 'common sense' but it is very rarely stated, and this is usually to the detriment of the lesbian couples. Most discussions are biased towards the heterosexual discussion of the menopause. Some people have reported research that suggests that lesbian women feel 'less regret' about the menopause. This could be because lesbian women may have fewer concerns about femininity, and of course, do not worry about being 'less attractive' to the opposite sex. Nevertheless, the menopause is a big change in any woman's life, and this applies to both lesbian and straight women. There may be a different emphasis, but the feelings and emotions of ending menstruation and moving to another phase of life cannot be dismissed lightly. This is a life-changing event with which every woman has to cope. There is no male equivalent (unless the andropause is considered equivalent, which is hard to rationalise), so in a male homosexual relationship, there is no such challenge. Male homosexuals also have the added advantage that, assuming they have sufficient money, a child is possible by egg donation and surrogacy at any time during their relationship.

PSYCHOLOGICAL STRENGTH

There is increasing evidence that lesbian women possess increased psychological strength when compared to straight women. This strength has been particularly noted in the context of aging where lesbian women seem to have a much brighter outlook on the process and consequences of aging. It is possible that lesbian women have a greater 'gender flexibility' when it comes to aging, which means that they may be less vulnerable to conform to 'traditional' feminine attributes of youth and fertility. This is because lesbian women have most likely defied the concept of 'femininity' for most of their lives and therefore do not see it as a 'loss'. There is a concept of 'relief to regret' regarding the ending of menstruation. Relief that the monthly pain, often pre-menstrual tension and bleeding, has gone but then the regret that their fertility is no longer present. This is a difficult and complex emotion to rationalise by all women regardless of their sexuality. A lesbian woman, for example, may have fully intended to have a child by artificial insemination, and an early or unexpected menopause will have completely and unexpectedly stopped these plans. This can be very traumatic. Equally a heterosexual woman could have had exactly the same plans and exactly the same outcome. Here the experience, regardless of sexuality, are equally traumatic, and it is difficult to see any differences between the lesbian and straight women.

Sexuality is a complex subject beyond the scope of this book. This chapter has aimed to try to raise awareness about the menopause and lesbian women. I hope that it is a success, but if not, at least I tried!

KEY POINTS OF CHAPTER 10

- The biological mechanism of the menopause in lesbian and heterosexual women is the same.
- Lesbian women may feel less 'listened to' or understood than their heterosexual peers. This may be a good area for healthcare professionals to consider and a possible area for better training and understanding.
- If a lesbian couple both experience the menopause at the same time, then this could be a very stressful situation. Help should be sought from family, friends, and of course, from healthcare professionals.
- Much more research is needed on the impact of the menopause on lesbian women. This will enable healthcare professionals to provide a better, supportive service to this extremely important group of patients.

CHAPTER 11

Advice to Menopausal Patients

(Keeping Positive)

Keep Buggering On!

Winston Churchill

INTRODUCTION

The whole of this book contains advice which may or may not be useful to the reader. I hope that it has at least been useful to at least some people. In such a complex and diverse subject as menopause, it is impossible to generalise or to say that any patient will 'definitely suffer' a range of symptoms. This means that it is equally difficult to give 'general' advice. Arguably the most important factor is education and understanding not only by the menopausal woman by women but also by the rest of society. There are approximately 1 billion women suffering perimenopause or menopausal symptoms at any time, making this education process enormous. In the UK alone, it is estimated that there are 13 million women suffering from perimenopase or menopause. The cost of providing adequate support and education from healthcare professionals in the UK alone would currently be prohibitive and, on a global scale, impossible.

There is actually very little formal teaching for women and girls about the menopause and even less for the general population on a global scale. The result of this is that the average woman has very little awareness and understanding of the menopause, which is a life-changing phase of female life. It must also be kept in mind that this menopausal life-changing event comes at approximately the time of half of the lifespan of a woman in developed countries. This leaves a very long time in which most women will have to cope with and live with menopause. This lack of understanding also extends to some healthcare professionals who do not receive adequate training to correctly and safely manage the menopause. This means that the onus is often on the woman undergoing the menopause to seek her own information and support. Our society is 'information heavy' because of the internet, which has transformed the availability of 'instant information' to everyone everyone. Sadly, all of the information on the internet is incorrect or valid and might even very often be deliberately misleading or harmful. This probl-

is worse in some internet 'chat rooms' where 'expert' opinions and 'advice' is especially erratic. If a woman seeks further information, then it is recommended to only look at professional websites, most of which are listed at the end of this book. Please do not get involved in any discussions or random 'advice' on the internet, as this may only increase your anxiety which is already at a heightened level.

It is interesting to note that the medical literature suggests that women *and* healthcare professionals may have an inadequate understanding of how the hormonal changes and the biological and psychological changes seen in the menopause relate to each other. This is particularly worrying if it truly applies to healthcare professionals. It is also true that most women do not understand the symptoms and consequences of the biological changes which take place during perimenopause, menopause and beyond. This may mean that some women may not seek the medical advice and support they so badly need.

THE FEMALE EXPERIENCE OF THE MENOPAUSE

It is known that in the UK education system, there is currently no formal education provided to girls about the menopause. This is in contrast to other diseases, such as sexually transmitted diseases and, more recently, COVID-19, where educational material is freely and widely available. This means that a high proportion of women fear the menopause because they have no basic understanding of what it means, what physical changes will happen to their bodies and perhaps most importantly, what psychological changes may occur, such as 'brain fog' and depression. In general terms, the media present a negative view of the menopause, but once again, this may be because of a simple lack of thorough research on their part. If clear and correct education relating to the menopause can be introduced in the UK (and in all other Countries), then when the perimenopause begins, then women will feel far less frightened and more in control of the situation. This should be a priority both for Government and Educationalists on a global scale. It should be remembered that not all of the symptoms which may appear around the time of the perimenopause will necessarily be caused by the perimenopause. Some could be totally unrelated, for example, aching legs after a long walk in perimenopausal women is not a symptom of perimenopause it is a symptom of tired muscles following a long walk. Some 'common sense' is needed here, but that common sense is harder to apply if the basic knowledge is either low or non-existent.

The basic symptoms of perimenopause and menopause have been described earlier in this book. What is clear is that any, or all, of these symptoms, can have a big effect on the quality of life of women, including their family life, sexual relations, and careers, if they are not managed properly. There is even sometimes the very unfortunate situation of the children of a perimenopausal woman undergoing puberty at the same time. The symptoms of puberty are similar to perimenopause, and the clashes resulting could make the perimenopausal woman and pubertal child feel worse. In such a situation, the perimenopausal woman and the pubertal child should seek medical advice and support, ideally together. This will help better understand both people involved and a much more relaxed family life.

Not surprisingly, most women suffer a loss of libido during the menopause, often related to vaginal dryness. This was covered earlier in this book in the symptoms of the menopause. This can cause relationship problems and can be an extremely damaging aspect of the menopause. Once again, medical advice and medication may be able to resolve this problem.

It is interesting that many women who are suffering perimenopausal symptoms do not realise that the symptoms they are suffering are due to the perimenopause. This causes considerable anxiety and stress, especially if the symptoms are typed into 'Dr Google', who may return with a horrific 'diagnosis' which just causes more stress and anxiety. It has also been reported that many women experience anger about how little they know about perimenopausal and menopausal symptoms. The healthcare systems are clearly letting these women down and action is needed to correct this very real problem. Some women even think that the symptoms of menopause cannot be treated. This means that they just 'suffer in silence', which should not be happening in the 21st Century. The current situation can only lead to the conclusion that at present, the quality of life of women is severely reduced by the perimenopause and menopause and that healthcare providers are in general terms, not prepared or educated to help these patients safely and effectively. This need not be the case, and I would urge all women, of all ages, to press for better education and support for the menopause. The resultant decrease in both physical and psychological suffering would be enormous and be a global benefit to society.

Female and Male Education

As in all aspects of life, education often increases the quality of life. If a person is well educated, then they are better placed to understand what is happening around them and make the best decisions to optimise their lifestyle and quality of life. This does not mean that everyone needs a Ph.D in Reproductive Physiology from Cambridge University to understand the menopause, far from it. The additional education needed to enable women, on a global scale, to better understand the menopause can be delivered at the school level in the same way that teenage girls receive education on menstruation, contraception and pregnancy. Adding some basic information on the menopause would be such an easy and effective thing to do, but at present, it sadly does not happen. It is unclear why this happens. This lack of early-age female education (in both developed and developing countries) means that almost all women are unprepared for the menopause, and when the menopause inevitably arises, most women are totally unprepared. There have been studies involving middle-aged women and the menopause who, when provided with the appropriate information, felt that they managed, understood and coped with the menopause much better than their contemporaries. The women in the study felt empowered from a practical and psychological point of view following their education sessions. This is a powerful and important observation. Many women may find that talking to their friends about the menopause is a useful route to 'self-help' since friends who are already going through the menopause will often have a much better insight into the experience than anyone else. Even very brief educational sessions, which could easily be delivered by primary healthcare providers, have been reported to be very valuable, informative and supportive to perimenopausal women.

It is equally important that men and boys are given education and information so they can better support their wives, partners and mothers going through menopause. In the UK, we are lucky enough to have the Personal, Social, Health and Economic (PHSE) Educational Curriculum. This is improving the general education for both sexes in the UK, but there still seem to be mysterious gaps regarding menopause.

Primary Care Health Providers

Primary care health providers, such as General Practitioners and Family Physicians, are the first point of contact for patients suffering from any disease or minor injury. This means that these healthcare professionals will be the first point of contact for a perimenopausal woman seeking initial help and advice with the symptoms she is suffering. The general lack of thorough education of primary healthcare providers on the perimenopause and menopause is a problem which needs to be accepted and resolved. This is for the sake (and even sanity) of perimenopausal/menopausal women. If primary healthcare professionals cannot attain these skills and knowledge, then the critical level of trust between primary healthcare professionals and patients will deteriorate. This can only result in the overall poor management of the perimenopause/menopause. At present, primary healthcare professionals seem to focus on hormonal levels in the blood and if these are within the accepted or expected range, then no further action is taken. The highly complex emotional and sexual experiences of perimenopausal and menopausal women are largely ignored by most primary healthcare professionals.

This may be because it is more difficult for primary healthcare workers to discuss these highly emotive subjects when compared to the relatively easy discussions surrounding physical symptoms. Despite this, it is the emotional, sexual and relationship matters which are potentially most damaging to the health and welfare of the perimenopausal/menopausal woman. This 'dismissive' care of perimenopausal/menopausal women requires recognition and correction as soon as possible to ensure that the best and most comprehensive care is offered to patients in great need.

The medical profession, in general terms, has started to recognise these shortfalls. In some countries, effective educational action is being taken, allowing healthcare professionals to fully and effectively, engage with perimenopausal/menopausal women. There are also some excellent supporters of menopausal women, such as the British Menopause Society (BMS), which has a wide range of meetings, lectures, exhibitions and courses which are extremely valuable to menopausal women. More such organisations are needed on a global scale. This has led to more pro-active and informed menopausal patients who may even attend healthcare professional consultations armed with the latest NICE guidelines on the menopause! This is 'patient power,' which I totally support, but it is a sad situation to be in. Patients should not have to instruct their healthcare professionals.

Hormone Replacement Therapy (HRT)

The subject of Hormone Replacement Therapy (HRT) is discussed in detail earlier in this book. In practice, some patients find it very difficult to be prescribed HRT and others find that HRT is prescribed far too freely. In a pressurised environment such as a General Practitioner or Family Physician surgery, there is often an emphasis on seeing patients as rapidly as possible (sometimes 10 minutes or less) and providing the 'cure', *i.e.*, a prescription, as quickly as possible. This is presumably what happens where HRT is prescribed too freely, where blood test results and a quick discussion with the patient results in a prescription for HRT. This high-pressure environment has not only been increased by COVID-19 but also by a general lack of primary healthcare physicians. This means that many surgeries are understaffed when it comes to physicians. More people are being trained, but it takes at least 7 years to train to be a fully qualified primary healthcare physician in the UK and a similar time-period in other countries. It is not a fast process, and it must not be rushed. This could result in a more general poor treatment of patients by primary care health professionals, which would be even more frightening. In the meantime, those primary healthcare physicians who are already working have to cope with enormous workloads (sometimes one GP or Family Physician can be responsible for thousands of patients), and their most precious commodity is time. The inevitable result is an illness in the primary healthcare physicians themselves or, worse still, the decision to take 'early retirement' to get away from the unbearable, never-ending workload. This situation is especially relevant in the NHS in the UK, which is perhaps why many menopause-specific private clinics are appearing in the UK where time, attention and clear advice are unlimited. Sadly, most people cannot afford to attend such private menopause clinics making state-provided healthcare the only option available. There is no simple answer to this conundrum.

In relation to HRT being difficult to get prescribed, this may be partially due to ineffective education of some healthcare physicians or more likely to be related to an inability to properly explain the alternative treatments open to menopausal women. Another aspect is that, at present, few healthcare professionals really understand the concept of personalised medicine. This is a concept where 'one-fits-all' cannot apply, and menopause is an excellent example of where 'one-fit--all' cannot and must never apply. Each patient needs careful assessment and time, thought and expertise put into which therapies (if any) are needed.

Anti-Depressants

Many perimenopausal women show signs of anxiety and depression, especially during the early phase of the perimenopause. This sadly leads to the over-prescribing of anti-depressants of various types to perimenopausal women. This is a case of treating the symptom, not the cause. In addition, anti-depressant treatments often have complex and unpleasant side effects, which can only complicate the already complex situation of the perimenopausal woman. An example of such a side effect of anti-depressants is a low libido which is also a symptom of the menopause. These two effects combined could result in zero libido with the resultant psychological and relationship damage. Once again, it is the responsibility of healthcare professionals to ensure that inappropriate prescribing of such medication is avoided. A popular 'catch-phrase' in modern medicine is that all treatment should be evidence-based. This means that diagnosis and treatment of any disease must always be made using all of the current evidence (e.g., medical journal publications) available. Evidence-based work is not the only way to practice medicine, but it is becoming clear that it might be the safest and most effective way. Evidence-based medicine is used in many instances, but in the case of the menopause, the 'evidence' seems to be ignored, and 'leaps-of-faith' or a 'hope for the best' attitude predominates. This is very sad for menopausal women.

Healthcare Professional Training

Healthcare professional training and Continued Professional Development (CPD) are essential for the safe and appropriate care of menopausal women and all patients. CPD ensures that healthcare professionals are operating using the latest ideas and technologies in their daily work. Despite this, CPD also takes time and money to deliver, so poorly funded hospitals (which are common in the UK) and over-worked and stressed healthcare professionals are by far not the perfect audience. Some people have suggested mandatory training in menopause for primary (GPs and Family Physicians) and secondary (hospital-based doctors and nurses) healthcare professionals. There is, after all, a wide range of mandatory training already in place for healthcare professionals, and adding menopause would, in theory, be easy. In practice, it would need funding, expertise and a will to provide this essential service. At present, it seems that none of these things are available in most countries, so progress on mandatory training in the menopause is likely to either be slow or non-existent. On the positive side, if patients start to complain about the lack of information and support for menopausal women, then the wheels might start to turn very slowly.

It should also be remembered that menopausal symptoms can cause a decrease in the quality of work carried out by the menopausal woman or indeed an increase in the work absence levels. This has a big effect on the efficiency of work and the economy, which with the simple interventions of good patient education and correct menopausal treatment, would be avoided.

KEY POINTS OF CHAPTER 11

- There is a lack of general education and understanding of the menopause on a global scale. Things are changing slowly in some countries, but there is much more to be done.
- Healthcare professionals may not be well-trained in how to diagnose and manage the menopause. This may be resolved by better focus in medical and nursing schools and by effective Continued Professional Development for all healthcare professionals.
- The menopause may involve extremely disrupting symptoms for some women. The level of support for such women needs improvement.
- Both female and male basic education about the menopause needs improvement.
- HRT and all other treatment options must be carefully monitored for ongoing safety and efficacy.

CHAPTER 12

A Final Thought

(Some Possible Conclusions)

Every thought you create in your mind creates your future.

Nitin Namdeo

INTRODUCTION

This book has been full of thoughts. Some of them were unpleasant, some of them possibly helpful, and some of them even possibly a bit frightening. My aim was to bring more clarity and understanding to the subject of the menopause, and I hope that I have made a useful, if modest, contribution. This book does not claim to be comprehensive, and it does not claim to present any 'magic bullet'. It does not offer, or even begin to offer, all of the solutions, but it might suggest a few possibilities. The real purpose of the book is to try to reduce anxiety and fear in menopausal women and to optimise the treatment and support they receive. This is a simple aim that will be extremely difficult to achieve. The philosophy of high-quality care applies to all other diseases, so why not the menopause? Is it money? Is it that only women are affected by the menopause? Would it be different if a man lost his fertility in his mid to late forties and developed severe stress and anxiety? Why do we live beyond our reproductive years? Is there 'life after the menopause?' I could pose these questions endlessly and still not come to a clear conclusion. I can conclude that menopause is a complex condition; it is currently poorly managed and supported in most countries, and we could certainly all do a lot better!

MENOPAUSE AND THE ANIMAL WORLD

In the human world, menopause is generally looked at as a negative thing because the menopausal woman can no longer reproduce. This is quite the opposite in some areas of the animal world. I am no David Attenborough, but I would like to give a few thoughts on the menopause in the animal world. It has long been known that fertility in animals decreases slowly as age increases; this is known as reproductive senescence. Despite this, in parts of the animal world, fertility does exist even at old age and death, but it is at a decreased level. It is very interesting note that humans are the only primates who do not die shortly after fertility ceases. This does not mean that humans simply have better healthcare and therefore survive post-fertility much better than animals. Studies of humans where healthcare is either very basic or sometimes non-existent still show that humans survive long after losing their fertility. It is, of course, very difficult (even impossible) to measure hormone status and ovulation in wild animals making it very difficult to properly understand the menopause process in wild animals.

When I was working on my 'A' level biology project I decided to study a very pretty and colourful (at least the male is colourful) type of freshwater tropical fish called guppies. These little fish can be seen at most aquaria and garden centres and have the distinction of being a group of fish that 'give birth' to live young (most fish lay eggs). I found this fascinating, leading partly to my life-long interest in science and medicine. It appears that female guppies do, in fact, go through a 'fish menopause' and lose their fertility for the last 14% of their lives. This is interesting, but it is still astonishing that these little female fish remain fertile for 86% of their lives, whereas female human remains fertile for approximately 50% of their lives. Similar observations have been made in other fish, mammals and invertebrates (animals without a back-bone). Despite all of this, there are still many unknowns. For example, some people think that the 'menopause'observation in guppies occurs only in relation to guppies in captivity. It is, of course, impossible to prove this hypothesis because studying guppies in the wild is currently impossible. Such are the problems that animal biologists face!

SLIGHTLY MORE CONVINCING EXAMPLES OF FEMALE ANIMAL MENOPAUSE

Whilst I am fond of guppies, there have been more convincing studies on larger aquatic mammals, such as whales which can be tagged and traced relatively easily. This is, nevertheless still a very specialised (and potentially dangerous to the scientists) branch of science. One thing to take into account is that observations such as these in the wild can never be 'controlled' in the way that laboratory observations can be 'controlled'. This means we do not know if any other extrinsic factors are contributing to what we see. For example, there could be unseen illness in the wild animal being studied, it may have been in a fight and have internal injuries, and it may be being poisoned by the many plastics and other toxins in our oceans. There are so many unknown variables that it is difficult to be fully confident in any data we collect. Despite all of this, it appears that female 'toothed whales' seem to live significantly after menopause. Equally, the magnificent killer whale seems to be fertile between the ages of 12 and 40 but has shown to be capable of living up to the age of 90. Similarly, short-finned pilot whales are fertile between the ages of 7 and 35 but very commonly live to age 60 and perhaps beyond.

At the other end of the scale, there is a tiny aphid with the Latin name *Quadrartus yoshinomiyai* in which post-menopausal females take up an important role in defending the colony. No one understands why this may be.

THE GRANDMOTHER HYPOTHESIS

On the face of it, the menopause seems to be a pretty silly thing to do from an evolutionary basis. Charles Darwin (who was based at Cambridge University) developed his Theory of Evolution through Natural Selection when travelling around the world on his ship, The Beagle, and observing the vast diversity of life on planet Earth. His basic conclusion was 'survival of the fittest', *i.e.*, if an animal is born with a feature that enhances survival (resulting from a slight change in genes), then this feature will be retained in future generations as an evolutionary advantage. The reason that menopause seems to go against the Theory of Evolution is that it would make sense for any individual creature (human or animal) to have a strategy to pass on their genes for as long as possible during their lifetime to optimise 'survival of the fittest'.

In order to try to explain this anomaly, it is necessary to use something which has become known as the 'Grandmother Hypothesis'. Please note that this is an 'hypothesis', not a 'theory', which means that it is currently just an idea that has yet to have formal proof. Nevertheless, the Grandmother Hypothesis proposes that females lose their fertility early to help their children and grandchildren reproduce. In the killer whale population, for example, there is some evidence to support this hypothesis because the older menopausal females have considerable knowledge built up over their years of experience, especially in where to find food, which they pass on to younger members of the pod. In this way, mothers increase the survival rate of their adult sons, which means more genes are passed on by fit adult males.

ELEPHANTS

Elephants are a fascinating species that are threatened with extinction in some parts of the world by the illegal trade in ivory. An adult elephant, weighing several tons, is often shot, and just the tusks are removed. The enormous carcass is just left to rot. This is an extremely sad example of human behaviour, but it is driven by the 'black market' value of ivory. Ivory cannot, in legal circumstances, be traded, but this still does not stop elephant poachers. A similar problem exists in rhinos where the horn is removed. Another example would be the removal of a sharks' fin and the rest of the dead animal discarded. We must stop this kind of destruction if our natural world is to survive. Humans must respect the other inhabitants of our planet.

Getting back to elephants, they stay together in a herd with a matriarch leader. Female elephants do not, as far as we know, go through the menopause. The difference in elephant society, as compared to an example from killer whales, is that in the elephant herd, males leave the herd when they are mature. This enhances genetic diversity within the herd, which is essential for biological fitness. Male elephants, mating with the matriarch, who would be their mother, would result in 'in-breeding' and severely reduce the genetic diversity of the herd.

Elephants have a complex society, and a dominant and extremely experienced female leads this society. Elephants are quite difficult to study in the wild, but as we learn more, we might even learn some useful lessons that could be applied to human society.

COMPETITION FOR FOOD AND RESOURCES

In the animal world, there is considerable competition for food and other resources such as minerals. The care and adequate feeding of their young are critical to their ongoing survival. If the young are poorly cared for or not fed properly, then the future of the species is in question.

In many animals, the 'hunters' are females, the most obvious being lions.

From a human viewpoint, most anthropologists agree that, in the early history of our species, daughters would move out of the birth family when they mature and go to live with a new family. In this way, the female would have no genetic relationship to her 'new' family until she had children within that family. As the same woman became older, she would become increasingly genetically related to her family as she had more children. Following the menopause, the same woman would be involved in raising the other children of her family, which is of course, genetically beneficial to her. If she were to continue having children (*i.e.*, not undergo the menopause) then the other children in the group would face greater competition for food and resources. This could be a catastrophic biological situation in very early human history who would have been hunter-gatherers and not organised farming would have existed. The menopause may therefore be an ancient biological process that ensured survival and genetic diversity in the early human race. Without this process, the human race might not have developed the way we see it today. In modern society, some may argue that the menopause is 'obsolete'. Nevertheless, menopause is built into female human physiology and is unlikely to change unless we choose to make direct interventions (such as the stem cell ideas earlier in this book).

CONCLUSION

This book has covered many complex ideas and concepts which I hope that I have explained clearly and in a readable style. There is a detailed glossary at the end of the book, which may help some readers to develop a better understanding. The book does not attempt to give any specific medical advice in any way, and it *must not* be interpreted in this way. It merely describes what is currently happening on the subject of menopause, the effect of menopause on our modern society, the symptoms of the menopause, the treatment of the menopause, and many other related and supportive pieces of information. If the book helps in the understanding of the menopause for both menopausal women and the rest of society, then my time has been well spent. In reality, this book is probably a 'drop in the ocean' in the overall debate, understanding and management of the menopause, but with the correct readership, debate and action, this 'little drip' might one day turn into an ocean.

KEY POINTS OF CHAPTER 12

- Female (and male) education and understanding about the menopause are essential. Without this then the menopause will continue to be a time of great fear and anxiety for billions of women and a time of uncertainty and worry for males.
- Unlike most other primates, female humans do not die shortly after fertility ceases.
- Fish such as female guppies seem to retain their fertility for approximately 86% of their lifespan.
- The rest of the animal world seems to have very different fertility patterns compared to humans, although of course, observations are extremely difficult in wild animals.
- The 'Grandmother Hypothesis' helps to explain how the menopause links in with the Theory of Evolution.
- Female elephants do not seem to undergo the menopause, and inbreeding in the group is avoided by the males leaving the herd when they are mature.
- It is possible that the human menopause may be a natural mechanism to prevent over-population resulting in limited amounts of food and resources being available. This concept applies to very early human history, where humans were hunter-gatherers making food and resources much more difficult to obtain. The introduction of agriculture in modern human civilisations means that food and resources are more readily available, which makes this concept unlikely to be applicable in the modern world.

USEFUL LINKS

https://aestheticmanagementpartners.com/exo-e/

https://assets.publishing.service.gov.uk/government/uploads/system/uploads/attachment_data/file/447673/motorcyclist-casualties-2013-data.pdf

https://www.atcm.co.uk/acupuncture-has-been-recommended-as-one-of-the-favourable-nonpharmacologic-treatments-for-low-back-pain-by-american-college-of-physicians#:~:text=Acupuncture%20as%20a%20treatment%20for%20lower%20back%20pain,Qaseem%20et%20al.%2C%202017%20made%20this%20recommendation.%20recommendation.

https://www.balance-menopause.com/

https://www.cellr4.org/article/2990 (The action of the QiLaser on VSEL stem cells).

https://www.cellr4.org/article/3201 (The mechanism of action of the QiLaser on VSEL stem cells using concepts from Quantum Physics).

https://www.cellr4.org/article/3280 (The use of QiLaser activated cord blood MSC and VSEL stem cells in the treatment of end-stage heart failure).

https://www.cellr4.org/article/3304 (A comprehensive review of the origins and properties of VSEL stem cells).

https://e-cigreviews.org.uk/vaping-side-effects/

https://www.cipd.co.uk/knowledge/culture/well-being/menopause/people-professiona-s-guidance

https://www.cipd.co.uk/Images/menopause-people-professionals-top-tips_tcm18-55428.pdf

https://www.cipd.co.uk/Images/menopause-guide-web_tcm18-55426.pdf

https://www.cipd.co.uk/Images/menopause-manifesto-roll-fold_tcm18-99975.pdf

https://www.daisynetwork.org/

https://www.healthline.com/nutrition/bmi-for-women

https://health.gov/our-work/nutrition-physical-activity/dietary-guidelines

https://www.gov.uk/government/publications/personal-social-health-and-econ-mic-education-pshe

https://healthtalk.org/menopause/overview

https://www.heart.org/en/healthy-living/healthy-lifestyle/quit-smoking-tobacco/5-steps-to-quit-smoking

https://menopausefriendly.co.uk/

https://www.nhs.uk/conditions/menopause/

https://www.nice.org.uk/guidance/ng23/ifp/chapter/About-this-information

https://nimh.org.uk/herbal-resources/herbal-support-for-menopause/

https://thebms.org.uk/

https://www.menopause.org/

https://www.menopause-exchange.co.uk/index.htm

https://www.nhs.uk/live-well/eat-well/how-to-eat-a-balanced-diet/eight-tips-for-healthy-eating/

https://qigenix.com/

https://www.samaritans.org/

https://www.whi.org/

https://www.youtube.com/watch?v=ahTQ6g7A5bM

https://www.youtube.com/watch?v=OUJ1Enik3yk

SUGGESTED FURTHER READING

http://www.eurekaselect.com/ebook_volume/2925

https://www.amazon.com/Menopause-Manifesto-Health-Facts-Feminism/dp/0806540664/ref=sr_1_1?crid=IOL3FYPVU9ZI&keywords=menopause&qid=1653571333&s=books&sprefix=menopause%2Cstripbooks-intl-ship%2C134&sr=1-1

https://www.amazon.com/Menopause-Your-Management-Rest-Life/dp/1732384862/ref=sr_1_3?crid=IOL3FYPVU9ZI&keywords=menopause&qid=1653571418&s=books&sprefix=menopause%2Cstripbooks-intl-ship%2C134&sr=1-3

GLOSSARY OF TERMS

Allogeneic: Cells or tissues which are genetically dissimilar and hence immunologically incompatible. These cells or tissues are always from individuals of the same species. Examples are heart and kidney transplants or allogeneic cell therapy such as a donor bone marrow transplant.

Antibodies: A blood protein produced in response to, and counteracting, a specific antigen. Antibodies combine chemically with substances which the body recognizes as alien, such as bacteria, viruses, and foreign substances in the blood.

Antigen: A toxin or other foreign substance which induces an immune response in the body, especially the production of antibodies.

Aphid: A small insect which feeds by sucking sap from plants; a blackfly or greenfly. Aphids reproduce rapidly, sometimes producing live young without mating, and large numbers can cause extensive damage to plants.

Autologous: Cells or tissues obtained from the same individual which may also be returned to that same individual. An example of this would be autologous QiLaser activated VSEL stem cell treatment as described in this book.

Autonomic Nervous System: The part of the nervous system responsible for control of the bodily functions not consciously directed, such as breathing, the heartbeat, and digestive processes.

Biological Age: This is a concept often used in current 'anti-aging' discussions to describe a shortfall between a population cohort average life expectancy and the perceived life expectancy of an individual of the same age. An example would be someone with a chronological age of 60 years and a biological age of 50 years. Such a person may be considered to have a greater likelihood of a longer 'healthspan'. This has yet to be proven. Many biomarkers are used to define biological age and they decline roughly linearly with age with a slope of <1% per annum. The general understanding of longevity and 'anti-aging' is in its' infancy (pardon the pun).

Biomarker: A naturally occurring molecule, gene, or characteristic by which a particular pathological or physiological process may be identified or defined.

Blood-Brain Barrier (BBB): This is layer of cells which keeps the brain safe from pathogens and other toxins or some medication. It is composed of brain cells and blood vessel cells. It plays a crucial role in protecting the most important organ in the body: the brain.

Blood-Follicle Barrier (BFB): The blood-follicle barrier (BFB) is one of the blood-tissue barriers in mammalian body found in developing follicles in the ovary. It protects the contents of the follicle which is follicular fluid and the developing human egg (proper term oocyte) from any possibly damaging molecules are toxins in the female body.

Blood Proteins: A broad term encompassing numerous proteins, including haemoglobin, albumin, globulins, the acute-phase proteins, transport molecules and many others.

Blood-Testes Barrier (BTB): The blood-testis barrier is a biological barrier separating the blood from the developing sperm in the testis. Damage to the BTB (chemical or physical) can result in damage to developing sperm and in turn to male infertility.

Carbohydrate: A large group of organic (carbon based) compounds occurring in foods and living tissues and including sugars, starch, and cellulose. Carbohydrates contain hydrogen and oxygen in the same ratio as water (2:1, i.e. H_2O) and typically can be broken down to release energy in the body.

Chronological Age: The number of years a person has lived, especially when used as a standard against which to measure behaviour and intelligence.

Class II HLA: These molecules, sometimes known as Class II MHC (Major Histocompatability Complex) are usually only found on immune cells which 'present' antigen to the immune system.

Clinical Trial: A scientifically controlled study of the safety and effectiveness of a therapeutic agent (such as a drug or vaccine or cell therapy) using consenting human subjects.

Continued Professional Development (CPD): The process of tracking and documenting the skills, knowledge and experience that a person gains, both formally and informally, as they work, beyond any initial training. The process of CPD keeps healthcare professionals 'up-t--date' in how they manage and treat patients.

Cosmeceutical: A cosmetic that has, or is claimed to have, medicinal properties.

Cytokines: Any of a number of molecules, such as interferon, interleukins, and growth factors, which are secreted by certain cells of the immune system and have an effect on other cells.

DNA: A self-replicating molecule that is present in nearly all living organisms as the main constituent of chromosomes. It is the carrier of genetic information (genes).

Electrolyte: The ionized or ionizable (e.g. sodium and potassium) constituents of a living cell, blood, or other organic matter.

Evidence Based: An approach to medicine, education, and other disciplines that emphasizes the practical application of the findings of the best available current research.

Exosomes: Tiny particles produced and released by most types of animal and plant cells. These exosomes are thought to be involved in 'cell-to cell' communication and cell regulation in normal and disease states.

Evolution: The process by which different kinds of living organism are believed to have developed from earlier forms during the history of the earth.

Fallopian Tubes: A pair of tubes that carry the egg from the ovary to the uterus (womb). Natural fertilisation of the egg takes place in the Fallopian tubes.

Follicle Stimulating Hormone (FSH): A hormone secreted by the pituitary gland (at the base of the brain) which promotes the formation of eggs or sperm.

Genes: A specific sequence of molecules in DNA and RNA that is located on a chromosome and that is the functional unit of inheritance controlling the transmission and expression of one or more traits. This is achieved by the gene by specifying the structure of a particular protein or controlling the function of other genetic material.

GP: General Practitioner or Family Physician (Primary Care Provider)

Growth Factors: A substance, such as a vitamin or hormone, which is required for the stimulation of growth in living cells.

Healthspan: The part of a person's life during which they are generally in good health

Hippocampus: Ridges on the base of the brain, thought to be the centre of emotion, memory, and the autonomic nervous system.

HLA: A set of genes on chromosome 6 in humans which direct the production of cell-surface proteins responsible for the regulation of the immune system.

HLA Typing: Tests to determine if a patient has antibodies against a potential donor's HLA. The presence of antibodies means that a particular graft will be quickly rejected. HLA typing can also be used to establish paternity and in forensic medicine.

Hypothesis: A supposition or proposed explanation made on the basis of limited evidence as a starting point for further investigation.

Idiopathic: Relating to or denoting any disease or condition which arises spontaneously or for which the cause is unknown.

Inbreeding: The breeding of closely related people or animals, especially over many generations.

Incontinence: The inability of the body to either control urination or defaecation.

Insomnia: Prolonged and usually abnormal inability to get enough sleep especially due to trouble falling asleep or staying asleep.

Introspective: Characterized by examination of one's own thoughts and feelings, thoughtfully reflective

Intravenous Line: A tube used for the administration of fluids into a vein by means of a steel needle and a plastic catheter. The plastic catheter remains in place in the vein during fluid administration.

Libido: Sexual drive.

Liposuction: The surgical collection of fat, usually from the abdomen, which can be used to prepare Mesenchymal Stem Cells (MSC).

Liquid Nitrogen: Liquid nitrogen is supercool at -196°C. It is used for cryopreservation, cryosurgery, and cryomedicine.

Low Fat: Denoting or relating to food or a diet that is low or relatively low in fat, especially saturated fat.

Matriarch: A woman (or another female animal e.g. an elephant) who is the head of a family, tribe or herd.

Menopause: The end of menstruation by a gradual process. Menopause is a natural life transition.

Mental Health Act 2007: This is an Act of the Parliament of the United Kingdom. It amended the Mental Health Act 1983 and the Mental Capacity Act 2005. It applies to people residing in England and Wales.

Mesenchymal Stem Cell (MSC): A stem cell found in connective tissue such as fat capable of producing connective tissue (tendons, ligaments), bone and fat cells.

Natural Selection: The process whereby organisms better adapted to their environment tend to survive and produce more offspring. The theory of its action was first fully expounded by Charles Darwin, and it is now regarded as be the main process that brings about evolution.

Neuroticism: A tendency toward anxiety, depression, self-doubt, and other negative feelings.

Neurotransmitter: A chemical substance which is released at the end of a nerve fibre by the arrival of a nerve impulse and, by diffusing across the synapse or junction, effects the transfer of the impulse to another nerve fibre, a muscle fibre, or some other structure.

Oestrogen: Any of a group of steroid hormones which promote the development and maintenance of female characteristics of the body. Such hormones are also produced artificially for use in oral contraceptives or to treat menopausal and menstrual disorders

Optimism: Hopefulness and confidence about the future or the success of something

Osteoporosis: The literal meaning of this term is 'porous bones'. It is caused by a loss of protein and minerals from the bone, especially calcium. Bone mass and therefore bone strength is reduced making the bones fragile and much more likely to break.

Perimenopause: Perimenopause means "around menopause" and refers to the time during which your body makes the natural transition to menopause, marking the end of the reproductive years. Perimenopause is also called the menopausal transition.

Personalised Medicine: A type of medical care in which treatment is customized for an individual patient

Pessimism: A tendency to see the worst aspect of things or believe that the worst will happen.

Placebo Controlled Clinical Trial: This is a clinical trial in which there are two (or more)

groups. One group gets the active treatment, the other gets the placebo. Everything else is held the same between the two groups, so that any difference in their outcome can be attributed to the active treatment.

Placebo: A substance that has no therapeutic effect, used as a control in testing new drugs and technologies.

Platelet: A small colourless disc-shaped cell fragment without a nucleus, found in large numbers in blood and involved in blood clotting

Platelet Rich Plasma (PRP): Blood that contains more platelets than normal. To create platelet-rich plasma, a blood sample is taken from the patient and placed into a device called a centrifuge that rapidly spins the blood, separating out the other components of the blood from the platelets and concentrating them within the plasma (the clear fluid component of the blood).

Pluripotent: The property of a stem cell capable of giving rise to all of the different cell types in the body, e.g. a VSEL stem cell.

Premature Menopause: Menopause in young women (aged less than 40) often with an unexplained cause

Primate: In zoology, any mammal of the group that includes lemurs, lorises, tarsiers, monkeys, apes and humans.

Premature Ovarian Insufficiency (POI): Primary ovarian insufficiency, also called premature ovarian failure, occurs when the ovaries stop functioning normally before age 40. When this happens, the ovaries do not produce normal amounts of the hormone oestrogen or release eggs regularly. This condition often leads to infertility.

Primary Healthcare Professional: These people work on caring for people rather than specific diseases. This means that professionals working in general practice are generalists, dealing with a broad range of physical, psychological and social problems, rather than specialists in a particular disease area.

Progesterone: A steroid hormone released by the ovary following ovulation that stimulates the uterus to prepare for pregnancy.

Progesterone Intolerance: Progesterone intolerance is when patients are particularly sensitive to the hormone progesterone or quite often the synthetic form, progestogen. The body reacts to the progesterone or progestogen, causing symptoms that can be similar to premenstrual syndrome.

Progestogen: A natural or synthetic steroid hormone, such as progesterone, that maintains pregnancy and prevents further ovulation during pregnancy.

Puberty: The period during which adolescents reach sexual maturity and become capable of reproduction

Qualitative: Relating to, measuring, or measured by the quality of something rather than its quantity.

Quantitative: Relating to, measuring, or measured by the quantity of something rather than its quality

Red Blood Cells: Any of the haemoglobin-containing cells that carry oxygen to the tissues and carbon dioxide away from the tissues. In mammals these are typically biconcave disk in shape, lack a nucleus and are formed from stem cells in the bone marrow.

Reproductive Senescence: This is an age-associated decline in reproductive performance, which often arises as a trade-off between current and future reproduction. Given that mortality is inevitable, increased allocation into current reproduction is favoured despite costs paid later in life.

RNA: Ribonucleic acid, is a nucleic acid present in all living cells. Its principal role is to act as a messenger carrying instructions from DNA for controlling the synthesis of proteins, although in some viruses RNA rather than DNA carries the genetic information.

Secondary Health Care Professional: This is the specialist treatment and support provided by doctors and other health professionals for patients who have been referred to them for specific expert care, most often provided in hospitals.

Sectioned: Being 'sectioned' means that a patient is kept in hospital under the Mental Health Act. There are different types of sections, each with different rules to keep you in hospital. The length of time that you can be kept in hospital depends on which section you are detained under.

Senescence: The condition or process of deterioration with age.

Sex Hormone: A hormone, such as oestrogen or testosterone, affecting sexual development or reproduction.

Sexually Transmitted Diseases: Sexually transmitted diseases (STDs) are infections transmitted from an infected person to an uninfected person through sexual contact. STDs can be caused by bacteria, viruses, or parasites. Examples include gonorrhoea, genital herpes, human papillomavirus infection, HIV/AIDS, chlamydia, and syphilis.

Spermatogonia: A stem cell in these testes which produces sperm.

Stem Cell Homing: In stem cell science, the word "homing" describes stem cells' ability to find their destination, or "niche." Identification of specific cues that steer stem cells to their niche and increase the efficiency of the homing process is an area of intense research.

Testosterone: A steroid hormone that stimulates development of male secondary sexual characteristics, produced mainly in the testes, but also in the ovaries and adrenal glands.

Type II Diabetes: This is a serious condition where the insulin made by the pancreas cannot work properly, or the pancreas cannot make enough insulin. This means that the blood glucose (sugar) levels keep rising if no treatment is started. It is common in obese patients.

Uterus: The organ in the lower body of a woman or female mammal where offspring are conceived and in which they gestate before birth; the womb.

Very Small Embryonic Like (VSEL) Stem Cell: These stem cells express several markers associated with a pluripotent state, including CXCR4, Oct-4, Nanog, SSEA-1, Rex-1, Rif-1, and give rise into cells from all three germ layers (the basic layers of the developing embryo).

Vitamin D: Any of a group of vitamins found in liver and fish oils, essential for the absorption of calcium and the prevention of rickets in children and bone disease in adults. They include calciferol (Vitamin D2) and cholecalciferol (Vitamin D3).

White Blood Cells: A type of blood cell that is made in the bone marrow and found in the blood and lymph tissue. White blood cells are part of the body's immune system. They help the body to fight infection and other diseases.

SUBJECT INDEX

A

Abdominal liposuction 40
Abnormal follicles 3
Abuse, homophobic 76
Acne 18, 20, 22
Age 34, 35, 37, 39, 47, 48, 50, 51, 52, 53, 72, 74, 86, 87, 88
  biological 47
  menopausal 53
  natural 51
Aging 47, 48, 78
Alcohol consumption 75
Alzheimer's disease 8, 11, 42, 55, 56
Andropause 35, 36, 37, 38, 49, 68, 69, 77
  symptoms of 37
Anger 5, 66, 81
  women experience 81
Antidepressants 28
Anxiety 4, 6, 7, 8, 10, 32, 52, 53, 54, 67, 68, 76, 77, 80, 81, 86
  levels 77
  treatment 32
Anxiousness 26, 28
Apheresis 42

B

Benefits 11, 12, 18, 19, 23, 24, 26, 27, 28, 29, 30, 31, 32, 33, 45, 81
  cell-based 45
  global 81
Bioidentical hormone preparations 29
Blood 18, 41, 42, 46
  brain barrier 41, 46
  clotting risk 18
  disorders 42
  testes barrier 41
Body 4, 22, 35, 36
  fat 36
  hair 35
  mass index (BMI) 22
  swelling 22

temperature 4
Bone 11, 40
  growth 11
  pelvic 40
Bone marrow 39, 40, 42, 44, 45, 46
  collection 40
  transplant 45, 46
Brain 1, 4, 6, 9, 47, 55, 56, 57. 58, 63, 80
  fog 4, 6, 9, 55, 56, 57, 58, 63, 80
  processes 1
  repair 47
Breast cancer 18, 20, 23, 38, 76
  developing 76
  risk of 20, 23
British menopause society (BMS) 83

C

Cancer 8, 19, 21, 23, 27, 51, 53, 56, 60, 63, 65, 75, 76
  ovarian 76
  uterine 19
Catastrophic biological situation 89
Cell(s) 32, 39, 44, 46
  oxygen-carrying 44
  therapy 46
  to-cell communication 32
  worn-out blood 39
  worn-out skin 39
Cervical cancer 76
Chemotherapy 3, 21, 52
Chinese medicine 31
Chlamydia 76
Clonidine 28, 29
Cognitive behavioural therapy (CBT) 32
Complementary therapies 29, 33
Constipation 18, 29
Coronary artery disease 51
COVID-19 1, 52, 64, 76, 80, 83
  pandemic 52, 64
  vaccine 1
Cramp 21

Cytokines 44, 45, 46, 47

D

Damage 23, 40, 42, 43, 45, 50
　joints 40
　physiological 43
Dehydration 56
Dementia 55, 56
Depression 2, 4, 5, 6, 8, 22, 26, 28, 53, 54, 55,
　　65, 70, 72, 77, 80
　developing 54
　severe 54, 65
　suffered 53
　symptoms of 6, 53
Diabetes 9, 37, 56
　type II 56
Diet 11, 21, 22, 26, 37, 56, 61
　healthy 11, 22, 37, 56
　high-carbohydrate 21
Digestive disorders 6
Diseases 11, 13, 15, 25, 26, 29, 30, 36, 37, 38,
　　40, 42, 43, 45, 47, 48, 50, 51, 55, 56, 59,
　　60, 70, 71, 75
　cardiovascular 36
　infectious 48
　inflammatory 40
　kidney 38
　life-threatening 47
　liver 38
　menopause-related 75
　neurodegenerative 11, 42, 55
　neurological 45
Dydrogesterone 17

E

Egg(s) 16, 34, 78, 87
　donation 78
　release 16
Electric fan 67
Electrolyte balance 31
Emotional 54
　instability 54
　support 54
Emotions 12, 53, 65, 66, 69, 75, 77
　amplified 12
Endocrinology 9, 11
End-stage heart failure 47

Erectile 36
　dysfunction 36
　problems 36
Events 2, 5, 10, 13, 22, 52, 54, 55, 57, 62, 68,
　　73, 77, 79
　community 57
　destructive life 13
　life-changing 52, 77, 79
　natural physiological 57
　transient 62

F

Facial hair 6
Fallopian tubes 51
Fat 40, 56
　belly 56
　tissue 40
Female 61, 87, 89
　animal menopause 87
　human physiology 89
　reproductive physiology 61
Fertilisation 34
Fish menopause 87
Follicle stimulating hormone (FSH) 15, 16, 17
FSH blood test 16
Function 34, 51
　ovarian 51

G

Genetic disorders 52
Glands, endocrine 9
Grandmother hypothesis 88, 90
Growth factors 44, 45, 46, 47

H

Hay fever 30
Headaches 6, 8, 22, 29
Health, cognitive 11
Healthcare 11, 13, 15, 16, 17, 18, 25, 27, 37,
　　38, 54, 75, 77, 81, 83, 84, 87
　systems 37, 75, 81
　workers 54, 75, 77
Heart 4, 30, 38, 45, 60, 63, 65
　attack 4, 65
　disease 4, 30, 38, 45, 60, 63, 65
Heterosexual relationships 76, 77

Hippocampus 56
Hormonal 4, 6, 15, 38, 80
    changes 4, 38, 80
    medication 15
    treatments 6
Hormone(s) 9, 10, 11, 13, 17, 18, 19, 20, 24,
        29, 31, 33, 35
    bioidentical 29
    levels 13
    progesterone 18
    natural 18, 33
Hormone testosterone 20, 35
    male 20
HPV infection 76
HRT 18, 19, 20, 23
    and breast cancer 23
    medication 18, 19, 23
    routine 20
    treatment 23
Human papilloma virus (HPV) 76
Hypertension 37
Hypothalamus 4
Hysterectomy 3

I

Infection, blood fight 1
Injuries 36, 39, 45, 87
    internal 87
    spinal cord 45
    surgical 39
Insomnia 4, 5, 23, 32, 66, 67
    treatment 32
Intruterine system 19

L

Lesbian 74, 75, 76, 77, 78
    menopausal women 74, 75, 76, 77
    relationships 74, 75, 76
    women 74, 75, 76, 77, 78
Lesbian couple 75, 77, 78
    menopausal 75
Leukaemia 42

M

Medication 15, 18, 19, 23, 24, 25, 28, 30, 64,
        70, 81, 84

anti-anxiety 70
    anti-histamine 30
    immune-suppressant 30
Medicine, personalised 20, 84
Medroxyprogesterone 17
Megakaryocyte 44
Menopausal 9, 10, 11, 12, 13, 16, 17, 20, 29,
        31, 33, 50, 51, 53, 54, 55, 57, 65, 66, 67,
        68, 69, 70, 73, 79, 81, 85
    depression 53, 54
    symptoms 11, 16, 17, 20, 29, 31, 33, 51,
        55, 79, 81, 85
    woman 9, 10, 12, 13, 50, 53, 57, 65, 66, 67,
        68, 69, 70, 73, 85
Menopause 10, 16, 87
    diagnostic kits 16
    process 10, 87
Mental health problems 32
Mesenchymal stem cells (MSC) 40, 41, 49
Methods of taking HRT 18
Monoamine oxidase inhibitors 30
Mood disorders 51
MSC technology 40

N

Natural 10, 13, 16, 25, 29, 37, 44
    ageing process 10, 37
    biological process 13, 25
    bodily process 13
    chemicals 44
    endorphins 29
    ovulation 16
Nausea 21, 28
Neurological trauma 45
Neuroticism 54, 55
Neurotransmitters 32
Nortryptyline 30

O

Obesity 6, 22, 26, 75
Oestrogen 3, 9, 7, 10, 16, 17, 18, 19, 20, 21,
        22, 28, 56, 68
    hormone 7, 68
    release 19
    vaginal 19, 20
Oestrogen gel 9, 10, 17, 19
    levels 9, 10

supplements 17
Oil, olive 56
Oophorectomy 3
Osteoporosis 7, 26, 28, 36, 51

P

Pain 6, 12, 13, 19, 22, 23, 28, 31
　abdominal 22, 28
Parkinson's disease 51
Perimenopause 1, 3, 4, 6, 7, 9, 13, 15, 17, 53,
　　54, 55, 79, 80, 81, 84
　suffering 79
　symptoms of 4, 6, 53, 80
Periods 2, 6, 15, 20, 53, 71
　final menstrual 53
　heavy 15
　irregular 2, 20
　long 6
　monthly 71
　regular 20
Peripheral blood stem cells (PBSC) 42
Physiotherapy 8
Plant 32, 33
　based exosomes 32, 33
　derived exosomes 33
　exosome therapy 33
Post-menopausal women 22, 28
Premature ovarian 3, 16, 44, 51, 52
　failure (POF) 3
　insufficiency (POI) 3, 16, 44, 51, 52
Pre-menstrual 18, 78
　syndrome 18
　tension 78
Pressure 28, 37, 66, 70, 71
　high blood 28, 37
Progestogen(s) 16, 17, 18, 19, 20, 21, 22, 28
　hormone 19
　intolerance 18
　micronised 17
　natural 18
　side effects of 22
　therapy 18
Prostate, enlarged 38

R

Radiation therapy 52

Relationship 6, 27, 31, 69, 70, 71, 75, 76, 77,
　　78, 89
　genetic 89
　homosexual 76
　male homosexual 77
Repair 33, 39, 40, 41, 42, 43, 46, 47
　burn 33
　cell-based ovarian 47
Reproductive 1, 34
　mechanisms 34
　organs 1

S

Selective serotonin reuptake inhibitors
　　(SSRIs) 28, 30
Serotonin-noradrenaline reuptake inhibitors
　　(SNRI) 28
Sex hormone(s) 35, 36
　bonding globulin (SHBG) 35
Sex therapist 52
Sexual 6, 27, 51, 52, 76
　dysfunction 51, 52
　health of lesbians 76
　intercourse 6, 27
　transmitted disease (STD) 76
Skin 18, 19, 33, 38, 39, 67
　oily 18
　patches 19, 38
　repairs 39
Sleep 4, 5, 8, 18, 23, 26, 30, 32, 56
　broken 5
　disruption 30
　nights 8
　quality of 23
Spermatogonia 34
Sperm production 34, 35
Stem cells 1, 39, 40, 41, 42, 44, 45, 46, 49
　bone marrow 46
　cost-effective 42
　embryonic 41, 42
　mobilised bone marrow 42
　peripheral blood 42
　stem cell of 42
Stress 3, 6, 7, 20, 27, 37, 56, 58, 61, 65, 73,
　　81, 86
　developed severe 86
　reduction activities 37
Stressed menopausal woman 65
Suffering perimenopausal symptoms 81

Suicidal thoughts 6, 65
Surgery 3, 8, 39, 52, 83
  abdominal 3
Survival 9, 10, 88, 89
  ongoing 89

T

Technology, potential anti-aging 33
Testosterone 10, 20, 35, 36, 37, 38
  bioavailable 35
  injections 38
  therapy 38
Tissue(s) 1, 35, 39, 40, 42, 43, 45, 46
  connective 40
  matching 45
  neuronal 46
  regenerate 39
  repairing 39
Toxins 1, 31, 87
Traditional chinese medicine 31
Tranylcypromine 30
Traumatic event 13
Treatment 6, 9, 11, 28, 29, 30, 31, 33, 37, 38,
    40, 41, 43, 44, 47, 49, 50, 51, 59, 60, 61,
    62, 70, 76, 84
  anti-depressant 84
  autologous 47
  cell-based 49, 50
  gel 38
  mainstream 49
  medical 60
  plant-derived 33
Tricyclic antidepressants 30

U

Urinary 6, 7
  infections 6
  problems 6, 7

V

Vaginal 6, 12, 19, 27, 28, 32, 68, 81
  discharge 28
  dryness 6, 12, 19, 27, 32, 68, 81
  dryness treatment 32
  lubricants 27

VSEL stem cells 41, 42, 43, 44, 45, 46, 47, 48,
    49
  derived 43
  technology 42
  treatment 43

W

Woman 3, 18, 71, 78, 83
  healthy 3
  heterosexual 78
  married 71
  perimenopausal/menopausal 83
  pre-menopausal 18